PRACTICE
MAKES
PERFECT™

Basic Math

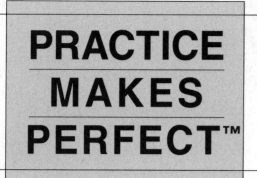

Basic Math

Second Edition

Carolyn C. Wheater

New York Chicago San Francisco Athens London Madrid Mexico City
Milan New Delhi Singapore Sydney Toronto

1 2 3 4 5 6 7 8 9   LHS   23 22 21 20 19 18

ISBN   978-1-260-13513-8
MHID   1-260-13513-6

e-ISBN   978-1-260-13514-5
e-MHID   1-260-13514-4

Interior design by Village Bookworks

McGraw-Hill products are available at special quantity discounts to use as premiums and sales promotions or for use in corporate training programs. To contact a representative, please e-mail us at bulksales@mcgraw-hill.com.

Also by Carolyn Wheater
Practice Makes Perfect: Algebra
Practice Makes Perfect: Geometry
Practice Makes Perfect: Trigonometry

Contents

Preface

In the middle of the 20th century, the terms *numeracy* and *innumeracy* entered into our vocabulary. The words were coined as parallels to *literacy* and *illiteracy*, the ability or inability to read and write at a level adequate to function in society. Numeracy was used to mean a numerical literacy, a competence with the basic mathematics that you need to function in the modern world.

When some people hear a phrase like *basic mathematics*, they think it just refers to simple arithmetic. That's not the case at all. A numerically literate person can do arithmetic, of course, but also understands our number system, has a sense of geometric and spatial relationships, and can intelligently analyze the flood of numerical information all around us. *Practice Makes Perfect: Basic Math* gives you the opportunity to strengthen your essential mathematical skills by review and practice.

Literate people not only *can* read; they *do* read. They practice the skills over and over again, day after day. Numeracy, mathematical literacy, also requires that you exercise the mental muscles that you use to calculate, to determine relationships, and to make sense of the data you encounter. Like any skill, basic mathematics needs to be practiced.

The exercises in this book are designed to help you acquire and develop the skills you need to be mathematically literate. With patience and practice, you'll find that you've assembled an impressive set of tools and that you're confident about your ability to use them properly. You must keep working at it, bit by bit. Be patient. You will make mistakes, but mistakes are one of the ways we learn, so welcome your mistakes. They'll decrease as you practice, because practice makes perfect.

Basic Math

Numbers and arithmetic

Just as every culture evolves its own language, different societies at different points in history have had different ways of writing and thinking about numbers. You've probably had some experience with Roman numerals, if only for dates on cornerstones, but for day-to-day use in arithmetic, you're far more comfortable with the numerals that come from Hindu and Arab mathematics. As opposed to the Mayan system of numeration based on 20, or some other ancient systems with other bases, our system is based on 10s, with the 10 digits 0, 1, 2, 3, 4, 5, 6, 7, 8, and 9 (unlike the Babylonian system with symbols only for 1 and 10). It's a **place value system**, meaning that the position of the digit within the number affects its value. The 7 in 57 is worth less than the 7 in 87,345, because they're in different places.

The families of numbers

The number system we use today didn't suddenly appear one day, fully formed. It developed and grew in response to people's need to count, to measure, and to evaluate different things. The **natural numbers**, or **counting numbers**, are just that: numbers used for counting. The natural numbers include 1, 2, 3, 4, and so on, and go on without end. You might notice there is no zero. If you don't have anything, you don't have to count it, so zero isn't a counting number. Once you start counting, however, you pretty quickly want a zero, so if you add zero to the natural numbers, you have what are called the **whole numbers**: 0, 1, 2, 3, 4, and so on.

Once you can count whole things, you find yourself wanting to talk about parts of things, so you start to need **fractions**. The set of numbers that mathematicians call the non-negative **rational numbers** include all numbers that can be written as fractions. That means it includes the whole numbers, because you can write them as fractions by putting them over a denominator of one, for example, 5 is $\frac{5}{1}$. The rational numbers also include every number that can be written in that form, like $\frac{1}{2}$ or $\frac{3}{7}$, or $\frac{114}{965}$, as long as the number on top is a whole number and the one on the bottom is a non-zero whole number.

Decimals, or decimal fractions, like 3.84 or 2.9, develop when you try to understand how fractions fit into our base 10 system, and then you discover that there are decimals that can't be written as fractions, and you realize that the rational numbers don't include everything. Numbers that can't be written as fractions are difficult to picture at first, but when you write them as decimals, the decimal goes on forever without repeating any sort of pattern. Those are called **irrational numbers**. Examples of irrational numbers include numbers like $\sqrt{2} \approx 1.414213562...$ and $\pi \approx 3.141592654...$ The dots tell you that the decimal keeps going forever.

Rational numbers, when changed to decimals, either stop, like $\dfrac{59}{4}=14\dfrac{3}{4}=14.75$, or if they go on forever, they repeat a pattern, like $\dfrac{1}{3}=0.333333...$ To show that a pattern repeats you can write a bar over the top. $\dfrac{1}{3}=0.\overline{3}$ and $\dfrac{16}{7}=2.\overline{285714}$.

Faced with these numbers that don't fit into the rational numbers, mathematicians expand their thinking to the **real numbers**, all the rational numbers and all the irrational ones as well. If you're wondering why they're called the real numbers—how could there be an unreal number?—you might be interested to know that there are numbers that mathematicians call imaginary numbers. You don't need to worry about those now.

EXERCISE

1·1

All of the numbers below are real numbers. Mark the sets of numbers they belong in.

	NATURAL OR COUNTING NUMBERS	WHOLE NUMBERS	RATIONAL NUMBERS	IRRATIONAL NUMBERS
1. 7	_____	_____	_____	_____
2. $\dfrac{3}{4}$	_____	_____	_____	_____
3. 1.9	_____	_____	_____	_____
4. 0	_____	_____	_____	_____
5. $4\dfrac{5}{7}$	_____	_____	_____	_____
6. 193	_____	_____	_____	_____
7. 18.328	_____	_____	_____	_____
8. 18.3284502193…	_____	_____	_____	_____
9. $6.\overline{29}$	_____	_____	_____	_____
10. $\dfrac{11}{23}$	_____	_____	_____	_____

Vocabulary and symbols

In talking about numbers, there will be some words, phrases, and symbols that you'll use. This is a good time to review them. You've already learned the vocabulary of sets of numbers: **natural** or **counting** numbers, **whole** numbers, **rational** numbers, **irrational** numbers, **real** numbers. You'll sometimes see those sets denoted by fancy letters, like $\mathbb{R}$ for the real numbers or $\mathbb{N}$ for the natural numbers.

If a number belongs in a certain set of numbers, it is an **element** or **member** of the set, and you'll sometimes see that written symbolically, using the symbol $\in$. For example, $4 \in \mathbb{N}$ means "four is an element of the natural numbers." The real numbers are a big set of numbers that contain the smaller sets like the natural numbers or the rational numbers. When a smaller set is completely contained in a bigger set, it's called a **subset**. The natural numbers are a subset of the real numbers. The symbolic form of that statement is $\mathbb{N} \subset \mathbb{R}$.

Basic arithmetic operations like addition and subtraction use standard symbols of $+$ and $-$, but multiplication and division have a little more variety. The multiplication problem "seven times eighteen" can be written as 7×18 or $7 \cdot 18$ or $7(18)$ or $(7)(18)$. The $\times$, the dot, and the parentheses all mean "multiply." (You'll sometimes see calculators and computers use $*$ as a multiplication sign, too.) To tell someone you want to divide 108 by 12, you can write $108 \div 12$ or $108/12$ or $\frac{108}{12}$ or $12\overline{)108}$. The last is usually used for long division and the others for simpler problems.

To tell what the result is, you use the equal sign, $=$, but if you want to say two things are not equal, you can write $\neq$. (Putting a slash through a sign is a common way to say "not." For example, you could write $0 \notin \mathbb{N}$ to say that zero is not part of the natural numbers.) If you want to be more specific than "not equal" and want to tell which number is larger, you can use the signs $>$ or $<$. The first says "is greater than," as in $19 > 5$, and the second says "is less than," as in $7 < 21$. The bigger number is at the bigger end of the symbol. Adding an extra line segment under the greater than sign, $\geq$, changes the meaning to "is greater than or equal to" and adding the extra segment to the less than sign, $\leq$, changes it to say "is less than or equal to."

When you start to do arithmetic, you encounter some new vocabulary. If you add, each of the numbers is an **addend** and the result is a **sum.** You subtract the **subtrahend** from the **minuend** to get a **difference.** (You don't hear *minuend* and *subtrahend* very much, but you should know the result of subtraction is a difference.) In a multiplication problem, each of the numbers is a **factor**, and the result is a **product.** For division, the **dividend** is divided by the **divisor** to produce a **quotient**, but there may be a **remainder.**

If the remainder of a division problem is zero, as is the case for $480 \div 16$, you'll hear that 16 is a **divisor** of 480, or that 16 is a **factor** of 480. If it sounds like a switch from division to multiplication, it is, because if $480 \div 16 = 30$ with no remainder, then $16 \times 30 = 480$. If the remainder is zero, the divisor is a **factor** of the dividend, and the dividend is a **multiple** of the divisor.

Whole numbers that are multiples of 2 are **even** numbers and whole numbers that are not even are **odd**. To find multiples of other numbers, you'll want to remember your "times tables," but you'll also want to know these divisibility rules.

2	Even numbers end in 0, 2, 4, 6, or 8.
3	If a number is a multiple of 3, the sum of its digits is a multiple of 3.
4	If a number is a multiple of 4, the last two digits are a multiple of 4.
5	Multiples of 5 end in 5 or 0.
6	Multiples of 6 are multiples of 2 and of 3.
9	If a number is a multiple of 9, the sum of its digits is a multiple of 9.
10	Multiples of 10 end in 0.

If a number has no factors except itself and one, the number is **prime**. If it has any other factors, it is **composite**. The only even prime number is 2, because all other even numbers are multiples of 2.

To add the same number several times, you can use multiplication as a shortcut. $5 + 5 + 5 + 5 + 5 + 5 = 6(5) = 30$. To multiply the same number several times, as in $2 \times 2 \times 2 \times 2 \times 2 \times 2$, you really can't find a shortcut for the work, but you can shorten the way it's written by using an **exponent**. $2 \times 2 \times 2 \times 2 \times 2 \times 2 \times 2 = 2^7$. The small, raised 7 is the **exponent**, and that tells you that you need to multiply 2, the **base**, seven times. The expression involving an exponent is commonly called a **power**, and the expression 2^7 is commonly read "two to the seventh power."

$$2 \times 2 \times 2 \times 2 \times 2 \times 2 \times 2 = 2^7 = 128.$$

The exponents you'll see most often are the second power, or **square**, and the third power, or **cube**. Those powers take their common names from the geometric figures square and cube. If you have a square in which each side measures 8 inches, the area of the square is 8^2. The volume of a cube that measures 7 by 7 by 7 is 7^3.

Almost every operation has an opposite, or **inverse**. The opposite of addition is subtraction and the inverse of multiplication is division. The opposite of exponentiation (the operation of raising a number to a power) is the process of taking a **root**. If $6^2 = 36$, then the **square root** of 36 is 6. The symbol for square root is called a **radical**. $\sqrt{36} = 6$. Every power has a corresponding root, but the basic symbol remains the radical. For roots other than square roots, you just add a little number, called an **index**, in the crook of the sign. If $4^3 = 64$, then $\sqrt[3]{64} = 4$. (More on roots and radicals can be found in Chapter 5.)

EXERCISE
1·2

For questions 1 through 10 fill in the blank with the correct word, phrase, or symbol.

1. To say "4.38 is a real number," you could write 4.38 _____ $\mathbb{R}$.

2. If the symbol for the irrational numbers is $\mathbb{Q}$, you can say "the irrational numbers are a subset of the real numbers" by writing _____ .

3. 6 _____ 4 = 24

4. 6 _____ 4 = 10

5. 6 _____ 4 = 2

6. 6 _____ 4 = 1.5

7. $\dfrac{16}{3} \in$ _____

8. $3 \times 3 \times 3 \times 3 =$ three to the _____ power

9. $\sqrt{9}$ _____ $\sqrt[3]{8}$

10. 7^2 _____ 5×10

For questions 11 through 15, complete each sentence with one of the symbols <, =, or > to make the sentence true.

11. The product of 8 and 3 _____ the quotient of 96 and 4.

12. The sum of 18 and 12 _____ the square of 4.

13. The difference between 100 and 80 _____ 5 squared.

14. Two to the third power _____ three squared.

15. The quotient of 84 and 6 _____ the product of 7 and 2.

For questions 16 through 25, label each statement true or false.

16. _____ 57 is a multiple of 3.

17. _____ 483 is a multiple of 9.

18. _____ 452 is a multiple of 6.

19. _____ 6^2 is a multiple of 9.

20. _____ $48 \div 4$ is a multiple of 6.

21. _____ 2,012 is a multiple of 4.

22. _____ 51 is prime.

23. _____ $68 \div 4$ is prime.

24. _____ $\sqrt{10}$ is a natural number.

Calculation

You may think that the growing presence of calculators and computers makes calculation, basic arithmetic, less important. While it's true that you're probably going to turn to a calculator for complicated operations or large numbers, you still need a command of how the calculation should be done. You want to be able to handle smaller, simpler calculations without scrambling for a calculator, and you want to be sure that you're getting the right results from your calculator. It's important that you not only know your basic addition and multiplication facts but also apply them in the correct sequence.

Order of operations

If a group of people looked at the problem $7 + 3 \times 5 - 6 \div 2 + 3^2 \cdot (5 + 1)$—admittedly, a complicated problem—and there were no rules for what to do first, second, and so on, each person could easily come up with a different answer. One person might decide to work right to left, while another might want to do operations one by one, and a third might jump on the "easy" pieces first. All of those are reasonable ideas, but they'll very likely all produce different answers. So how do you decide who's right?

The **order of operations** is the set of rules mathematicians have agreed to follow in evaluating expressions that have more than one operation. The order of operations says:

1. Do everything inside **parentheses** or other grouping symbols. If you have parentheses inside parentheses, work from the inside out.

2. Evaluate any powers, that is, numbers with **exponents**.

3. Do all **multiplication** and **division**, as you encounter it, from left to right.

4. Do all **addition** and **subtraction**, as you encounter it, from left to right.

People often use a memory device to remember the order of operations. Parentheses, Exponents, Multiplication and Division, Addition and Subtraction can be abbreviated as PEMDAS, which you can remember just like that or by a phrase like Please Excuse My Dear Aunt Sally.

Following those rules to evaluate $7+3\times5-6\div2+3^2\cdot(5+1)$, you first add the $5+1$ in the parentheses. Once you take care of what's inside the parentheses, you can drop the parentheses.

$$7+3\times5-6\div2+3^2\cdot(5+1)=7+3\times5-6\div2+3^2\cdot6$$

Exponents are next, if you have any, and this expression does. $3^2=9$, so

$$\begin{aligned}7+3\times5-6\div2+3^2\cdot(5+1)&=7+3\times5-6\div2+3^2\cdot6\\&=7+3\times5-6\div2+9\cdot6\end{aligned}$$

Multiplication and division have equal priority, so move left to right and do them as you find them. Jumping over a division to do all the multiplication first isn't necessary and may cause errors.

$$\begin{aligned}7+3\times5-6\div2+3^2\cdot(5+1)&=7+3\times5-6\div2+3^2\cdot6\\&=7+3\times5-6\div2+9\cdot6\\&=7+15-3+54\end{aligned}$$

Finally, start back at the left again and work to the right, doing addition or subtraction as you go.

$$\begin{aligned}7+3\times5-6\div2+3^2\cdot(5+1)&=7+3\times5-6\div2+3^2\cdot6\\&=7+3\times5-6\div2+9\cdot6\\&=7+15-3+54\\&=22-3+54\\&=19+54\\&=73\end{aligned}$$

EXERCISE

1·3

Simplify each expression, following the order of operations.

1. $35-17+22$

2. $505-100\times3$

3. $125-18\div6\times21+7$

4. $13\times20+35+(48-15)\div11$

5. $20\times3\div4+23-17+83$

6. $234\div6\times2+103$

7. $(200+300\times25\div15)\div10$

8. $2\times3+5-25\times4\div10$

9. $2\times(3+5)-25\times4\div10$

10. $5^2\div(12-7)+8\times3^2$

Integers

Your first experiences with arithmetic used whole numbers, and you learned two things very quickly. You can't subtract a bigger number from a smaller one, and you can't divide a smaller number by a bigger one. Your early encounters with $7-12$ and $3\div5$ went in the impossible column.

Soon, however, you learned about fractions and decimals, and $3\div5$ was just $\dfrac{3}{5}$ or 0.6. When your mathematical world expanded enough to deal with $7-12$, you discovered the set of integers.

The **integers** let you deal with ideas like owing and having, above ground and below ground, losing and gaining. They introduce the idea of numbers less than zero, each carrying its signature negative sign in front. The idea starts with the notion of an opposite. You've thought about addition and subtraction as opposite operations, but now the idea is that, for the operation of addition, every natural number has an opposite, and when you add a number and its opposite, the result is zero. So, $4 + -4 = 0$ and $18 + -18 = 0$. You're probably thinking that looks an awful lot like subtraction, and it does, because mathematicians define subtraction as adding the opposite. To subtract $38 - 23$, add 38 and the opposite of 23, $38 + -23$.

The integers are a set of numbers that include all the natural numbers and their opposites and zero. The natural numbers, the positive integers, are sometimes written with a plus sign in front but more often without. The negatives will always have a negative sign in front. It's hard to write out the set, but it looks something like this: $\{...,-4,-3,-2,-1,0,1,2,3,4,...\}$. Of course, once you start talking about opposites, it's easy to extend the idea beyond the natural numbers to the rational numbers and irrationals, and then you realize that the real numbers include an infinite number of positive numbers, as well as all their opposites, and zero. The positive numbers, greater than zero, and the negative numbers, less than zero, reflect one another with zero as the mirror.

Absolute value

The **absolute value** of a number is a measurement of how far from zero that number is. Think of the numbers arranged on a line, with zero dividing the positives from the negatives, and the integers evenly spaced along the line. Positives go to the right, negatives to the left, and the positives and negatives are mirror images of one another. Just imagine the integers for now, but remember that all the fractions and decimals, and all the irrationals, are in those gaps in between integers. This real number line is shown in Figure 1-1.

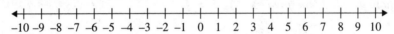

Figure 1-1

Now the number 7 is seven units away from zero—to the right of zero, but for absolute value, you only care about how far, not which direction. The number -5 is five units away from zero. It's five units to the left, but again, direction doesn't matter. The absolute value of 7, written $|7|$, is 7. The absolute value of -5, $|-5|$, is 5.

You can think of the absolute value of a number as the "number" part without the sign. You can think of it as distance without regard to direction. Or you can think of it as the distance of the number from zero. Whichever one makes sense to you, the idea of absolute value will help you understand operations with integers.

Addition

If you think about a positive number as a gain and a negative number as a loss, it's easy to see that a gain plus a gain is a gain, and a loss plus a loss is a loss. In other words, a positive number plus a positive number is a positive number, and a negative number plus a negative number is a negative number. What about a gain and a loss? Your first thought was probably "how big?" How big a gain and how big a loss? Do they cancel each other out? Did you win $1 and lose $100? That's a loss. But if you won $100 and lost $1, that's a win. When you're adding positive and negative numbers, the absolute values of the numbers are key to the answer.

To add integers with the same sign, add the absolute values and keep the sign.

$$4 + 9 = 13$$
$$-5 + -11 = -16$$

To add integers with different signs, subtract the absolute values and take the sign of the number with the larger absolute value.

$$19 + -8 = 11$$

Winning 19 and losing 8 means winning 11.

$$15 + -40 = -25$$

Winning 15 and losing 40 means losing 25.

EXERCISE
1·4

For questions 1 through 5, find the absolute value of each number.

1. -4

2. $+3$

3. 0

4. 178

5. -192

For questions 6 through 15, add the integers.

6. $-8 + 9$

7. $3 + -5$

8. $-7 + 11$

9. $12 + -9$

10. $-19 + 25$

11. $33 + -47$

12. $-87 + -91$

13. $44 + 35$

14. $-101 + 110$

15. $93 + -104$

Subtraction

When you're asked to subtract integers, don't. Remember that subtraction is defined as adding the opposite, so when you're looking at a subtraction problem, you can rewrite it as an addition problem. The subtraction problem $8 - 4$ can become $8 + -4$, and $-6 - (-3)$ can become $-6 + 3$. Notice that the first number doesn't change. The operation changes to addition, and the last number flips its sign. You can remember the pattern as *keep, change, change*.

Don't subtract. Change the sign of the second number and add. Follow the rules for addition.

$$\underset{first}{-14} - \underset{second}{(-6)} = \underset{keep}{-14} \underset{change}{+} \underset{change}{6} = -8$$

Perform each subtraction.

1. $19 - (-8)$

2. $-27 - 14$

3. $-35 - (-29)$

4. $42 - (-37)$

5. $-56 - 44$

6. $62 - 28$

7. $71 - (-35)$

8. $-84 - 70$

9. $-90 - (-87)$

10. $107 - 112$

Multiplication and division

Multiplication is actually repeated addition. When you write 3×5, you're actually saying "add three fives" (or "add five threes"). Adding a positive number several times will give you a positive result, so $3 \times 5 = 15$. Adding a negative number several times will give you a negative result, so a problem like 4×-8 will give you a negative answer. 4×-8 means $-8 + -8 + -8 + -8$ and that's -32. The tough one to think about is a negative times a negative, but if you remember that those negative signs are telling you "the opposite of" and you know that $4 \times -8 = -32$, then putting in another negative, as in -4×-8, should say "the opposite of 4×-8." So $-4 \times -8 = 32$.

When you're multiplying or dividing, the rules for signs are exactly the same. If the signs are the same, the result is positive.

$$7 \times 9 = 63$$
$$-6 \times -9 = 54$$

It doesn't matter if the numbers are both positive or if they're both negative, as long as they're the same sign.

$$-24 \div -8 = 3$$
$$35 \div 7 = 5$$

If the signs are different, the result is negative.

$$-14 \times 3 = -42 \qquad -100 \div 25 = -4$$
$$7 \times -12 = -84 \qquad 81 \div -3 = -27$$

Powers

When you write a power, such as 2^5, you're saying "multiply 2 five times." $2^5 = 2 \times 2 \times 2 \times 2 \times 2 = 32$. When a positive number is raised to any power, the result is positive.

When you write powers with bases that are negative numbers, there are a couple of things to watch. The first has to do with the writing itself. If you want to raise a negative number to a power, put the negative number in parentheses and place the exponent outside the parentheses, like this: $(-3)^2$. This tells your reader that you want to use the entire number -3 as a factor twice. $(-3)^2 = (-3) \times (-3) = 9$. If you don't use parentheses and you write -3^2, your reader will see that as "the opposite of 3^2," or -9.

The other thing to think about is the sign of the result. A negative times a negative is positive, so squaring a negative number gives you a positive result, but what if you raise a negative number to a larger power? Think about $(-2)^5$.

$$(-2)^5 = \underbrace{(-2)\times(-2)}_{\text{negative} \times \text{negative} = \text{positive}} \times(-2)\times(-2)\times(-2)$$

$$= \underbrace{4\times(-2)}_{\text{positive} \times \text{negative} = \text{negative}} \times(-2)\times(-2)$$

$$= \underbrace{-8\times(-2)}_{\text{negative} \times \text{negative} = \text{positive}} \times(-2)$$

$$= \underbrace{16\times(-2)}_{\text{positive} \times \text{negative} = \text{negative}}$$

$$= -32$$

Each time you multiply by -2, the sign changes. If the number of negative signs is even, the result will be positive, but if you have an odd number of negative signs, the result will be negative.

When a negative number is raised to an even power, the result is positive.

When a negative number is raised to an odd power, the result is negative.

EXERCISE 1·6

For questions 1 through 15, perform each multiplication or division, as indicated.

1. 18×2

2. $-5 \div -1$

3. -7×3

4. $50 \div -5$

5. 6×-4

6. $-12 \div -3$

7. -13×-5

8. $49 \div -7$

9. 11×-3

10. $-20 \div 4$

11. -34×3

12. $110 \div -55$

13. 9×-8

14. $-42 \div 7$

15. -6×12

For questions 16 through 20, evaluate each power.

16. $(-3)^2$

17. 2^3

18. $(-2)^3$

19. -5^2

20. $(-1)^4$

Factors and primes

Did you learn your "gozintas" in grade school? What's a gozinta, you ask? You know, four gozinta twelve three times. Relationships between numbers are often defined by division. Use half as much sugar as lemon juice in your lemonade, or mix oil and gas in a 1 to 50 ratio. Of course, division is the opposite of multiplication, but it's so closely connected to multiplication that you can't really talk about one without the other.

Prime or composite

If a number has no factors except itself and 1, the number is prime. If it has any other factors, it is composite. The only even prime number is 2, because any other even number is a multiple of 2 and so has 2 and some other number as a factor pair. The rest of the prime numbers are odd, but not all odd numbers are prime. The number 15, for example, is composite because, besides 1×15, it could also be expressed as 3×5. Because it has factors other than 15 and 1, 15 is composite. On the other hand, the number 19 has no factors except 19 and 1, so 19 is prime.

When you're dealing with smaller numbers like 27 or 31, it's not too hard to test all the possibilities. You can figure out pretty quickly that 27 is divisible by 3, so 27 is composite. It will take a little more work to determine that 31 is prime. It's not even, so it's not divisible by 2 (or 4 or 6 or any even number) and its digits add to 4, so it's not divisible by 3 or 9. It doesn't end in 5 or 0, so it's not divisible by 5 or 10 or any of their multiples. You'll need to try 7, but $31 \div 7$ leaves a remainder of 3, so 7 is not a factor. $31 \div 11$ and $31 \div 13$ also have remainders, so neither 11 nor 13 is a factor. You could keep trying 17, 19, 23, and 29, the other primes less than 31, but you don't actually have to go that far. You only have to try divisors up to the square root of 31. The square root of 31 is larger than 5 (because $5^2 = 25$) and smaller than 6 (because $6^2 = 36$), so you only have to try as far as 6. If you haven't found a prime factor of 31 less than $\sqrt{31}$, you won't find one greater than $\sqrt{31}$. Once you've checked 5 you can stop looking.

Label each number prime or composite.

1. 17
2. 29
3. 111
4. 239
5. 517

6. 613
7. 852
8. 741
9. 912
10. 2,209

Prime factorization

There are times when you're working with large numbers and realize it would be easier to accomplish your task if the number were expressed as the product of smaller numbers. (Simplifying fractions is one such situation.) It's often convenient to express the numbers as a product of prime factors. While a large number, such as 840, could be expressed in factored form many different ways—84×10, 7×120, 24×35, and others—its prime factorization is unique.

To find the prime factorization efficiently, you'll need to find factors of the number, and then possibly factors of the factors. It's a good idea to review the divisibility tests:

2 Even numbers end in 0, 2, 4, 6, or 8.
3 If a number is a multiple of 3, the sum of its digits is a multiple of 3.
4 If a number is a multiple of 4, the last two digits are a multiple of 4.
5 Multiples of 5 end in 5 or 0.
6 Multiples of 6 are multiples of 2 and of 3.
9 If a number is a multiple of 9, the sum of its digits is a multiple of 9.
10 Multiples of 10 end in 0.

To find the prime factorization of a number, you can use either of two strategies. In the first strategy, start with small primes like 2, 3, and 5 and divide your number by those. To find the prime factorization of 9,240, notice that 9,240 is divisible by 2, so $9,240 = 2 \times 4,620$. But 4,620 is also divisible by 2, so $9,240 = 2 \times 2 \times 2,310$. Keep dividing by 2 as long as you can.

$$9,240 = 2 \times 2 \times 2 \times 1,155$$

When the last number is no longer divisible by 2, check to see if it's divisible by 3. $1 + 1 + 5 + 5 = 12$, which is a multiple of 3, so 1,155 is divisible by 3. Divide by 3 as many times as possible (in this case, just once).

$$9,240 = 2 \times 2 \times 2 \times 3 \times 385$$

Move on to 5: 385 is a multiple of 5.

$$9,240 = 2 \times 2 \times 2 \times 3 \times 5 \times 77$$

Keep trying primes until all the numbers are prime.

$$9,240 = 2 \times 2 \times 2 \times 3 \times 5 \times 7 \times 11$$

Exponents can be used to write the prime factorization more compactly.

$$9{,}240 = 2^3 \times 3 \times 5 \times 7 \times 11$$

The other strategy is commonly called a factor tree. Start with your number, say 420, and find any pair of factors for it.

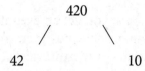

If the factors are not prime, find factors of each of them.

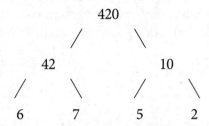

Repeat with any composite numbers, until you've reduced everything to primes.

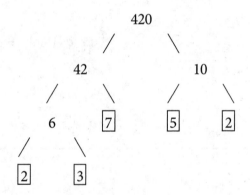

The boxed numbers are prime, so the prime factorization of 420 is $2 \times 3 \times 7 \times 5 \times 2$, or $2^2 \times 3 \times 5 \times 7$.

Find the prime factorization of each number.

1. 48
2. 92
3. 120
4. 168
5. 143

6. 300
7. 189
8. 364
9. 1,944
10. 5,225

Even or odd

Whole numbers that are multiples of 2 are even numbers and whole numbers that are not even are odd. Even numbers can be represented symbolically as $2n$ where the n stands for some whole number. Any odd number is one more than some even number and one less than some other even number, so odd numbers can be represented as $2n + 1$ or $2n - 1$. The value of n will differ depending on which format you choose.

When you add two even numbers, you get an even number, as when you add $6 + 8$ and get 14, all even. When you add two odd numbers, however, you get an even number, as you can see if you look at $5 + 7 = 12$. That may seem counterintuitive at first, but odds are evens plus 1, so even $+ 1 +$ even $+ 1 =$ even $+$ even $+ 1 + 1 =$ even $+ 2$, which is even. When you add an even and an odd, you get an odd.

When you multiply two even numbers, like 12 and 20, you get an even number, in this case, 240. When you multiply two odd numbers, you get an odd number. For example, $7 \times 11 = 77$. When you multiply an even and an odd, you get an even, as when you multiply $8 \times 15 = 120$.

EXERCISE
2·3

If a and b are even numbers and x and y are odd numbers, determine whether each of the following expressions would represent an odd or an even number.

1. $a + b$

2. $x + y$

3. $a + x$

4. $a \cdot b + y$

5. $y(a + x)$

6. $a \cdot x + b \cdot y$

7. $a - b$ (Think of $a - b$ as $a + -b$.)

8. $x - y$

9. $a \cdot y - b \cdot x$

10. $a^2 + b \cdot y - x^2$

Factors, divisors, and multiples

If the remainder of a division problem is zero, as is the case for $480 \div 16$, you say that 16 is a **divisor** of 480, or that 16 is a **factor** of 480. If that sounds like a mix-up of division and multiplication, remember that multiplication and division are inverse operations, but because of that, they're closely connected. If $480 \div 16 = 30$ with no remainder, then $16 \times 30 = 480$. If the remainder of the division is zero, the **divisor** is a **factor** of the dividend, and the dividend is a **multiple** of the divisor.

When you learned to divide, you probably learned to check your answer by multiplying. Is $489 \div 27 = 18$? You'd check by multiplying 27×18 to see if you get 489. If you do that multiplication, you'll find out that $27 \times 18 = 486$, so $489 \div 27 \neq 18$, at least not exactly. $489 \div 27 = 18$ with a remainder of 3, or, put another way, $489 = 27 \times 18 + 3$.

For questions 1 through 5, label each statement true or false.

1. 9 is a factor of 89.

2. 6 is a factor of 84.

3. 4 is a factor of 8,710.

4. 5 is a factor of 790.

5. 11 is a factor of 209.

Questions 6 through 10 present you with division problems. Perform each division and write the result in the form dividend = divisor × quotient + remainder.

6. $67 \div 11$

7. $83 \div 7$

8. $194 \div 13$

9. $1,197 \div 3$

10. $401 \div 8$

Greatest common factor

The numbers 39 and 51 both have a factor of 3 in their prime factorization. $39 = 3 \times 13$ and $51 = 3 \times 17$. The number 3 is a common factor of 39 and 51. The numbers 48 and 64 have several common factors: 2, 4, and 8. The largest of those, 8, is called the **greatest common factor**. Two numbers that have no common factors other than 1, numbers whose greatest common factor is 1, are **relatively prime**.

Finding the greatest common factor, or GCF, for large numbers is often easier if you first find the prime factorization of each number. If you're looking for the greatest common factor of 525 and 135, start with the prime factorization of each.

$$525 = 3 \times 5 \times 5 \times 7$$
$$135 = 3 \times 3 \times 3 \times 5$$

Notice which factors appear in both factorizations: one 3 and one 5. The GCF is 3×5, or 15.

To find the greatest common factor for 86,625 and 3,900, begin with the prime factorization of each number.

$$86,625 = 3 \times 3 \times 5 \times 5 \times 5 \times 7 \times 11$$
$$3,900 = 2 \times 2 \times 3 \times 5 \times 5 \times 13$$

The factorizations have one 3 in common and two 5s in common, so the greatest common factor is the product of $3 \times 5 \times 5$, or 75.

Find the greatest common factor of each pair of numbers.

1. 35 and 42

2. 48 and 60

3. 64 and 96

4. 65 and 117

5. 93 and 98

6. 98 and 105

7. 105 and 495

8. 121 and 169

9. 144 and 342

10. 275 and 2,125

Least common multiple

While the greatest common factor is the largest number that divides two or more numbers without remainders, the least common multiple is the smallest number that two or more numbers will divide. The least common multiple of 3 and 5 is 15, because 15 is the smallest number divisible by both 3 and 5. The least common multiple of 8 and 12 is 24 because 24 is a multiple of 8 (8×3) and a multiple of 12 (12×2) and it is the smallest number that is a multiple of both.

To find the least common multiple, or LCM, of two numbers, begin with the prime factorization of each number. If you want the least common multiple of 28 and 42, factor each one.

$$28 = 2 \times 2 \times 7$$
$$42 = 2 \times 3 \times 7$$

Notice what the GCF is. In this example, it's 2×7, or 14. The least common multiple starts there, but also includes other factors. If you strike out the 2 and the 7 that are common to the two factorizations, any factors left are also factors of the LCM.

$$28 = \cancel{2} \times 2 \times \cancel{7}$$
$$42 = \cancel{2} \times 3 \times \cancel{7}$$
$$LCM = \underbrace{2 \times 7}_{GCF} \times \underbrace{2 \times 3}_{remaining\ factors}$$

The LCM of 28 and 42 is 84.

If two numbers are relatively prime, their least common multiple is their product.

EXERCISE 2·6

Find the least common multiple of each pair of numbers.

1. 14 and 24

2. 34 and 85

3. 33 and 55

4. 90 and 72

5. 36 and 27

6. 35 and 42

7. 88 and 110

8. 48 and 60

9. 175 and 140

10. 64 and 96

Fractions

Once you have the arithmetic of whole numbers and integers under control, the next move is to master the arithmetic of fractions, whether common fractions, decimal fractions, or percents.

Fractions

Fractions are rational numbers in their most basic form. Each fraction is a division statement. The division symbol (÷) is a picture of a fraction, with the dots taking the place of the numbers. The top number, the dividend of the division problem, is the **numerator**, and the bottom number, the divisor, is the **denominator**. A fraction with a numerator of 1 is called a **unit fraction**.

Denominator comes from the same root as denomination. The denominator tells you what type of fraction you have, and is based upon the number of parts into which the whole was divided. If a pie is cut into eight equal pieces, each piece is one eighth, or $\frac{1}{8}$. When the same pie is divided into six pieces, each piece is one sixth of the whole, or $\frac{1}{6}$. The numerator tells you the number of those parts that you have. If the pie is divided into eighths and you eat three of those pieces, you've eaten three eighths, or $\frac{3}{8}$, of the pie.

Simplifying fractions

Just as both $\frac{6}{6}$ and $\frac{8}{8}$ represent a whole, there are many fractions that look different, but are equivalent. They name the same amount. If you cut a whole into two equal parts, each is $\frac{1}{2}$, but if you cut the same whole into eight equal parts, $\frac{4}{8}$ would describe the same part as the $\frac{1}{2}$.

$$\frac{4}{8} = \frac{1}{2}.$$

Since you don't want to start cutting things up every time you want to know if two fractions are equivalent, you need another tactic. You want to put each fraction in its simplest form. Once both are in their simplest forms, you can see if they're equivalent. The key to getting them into their simplest form is disguised ones.

A fraction in which the numerator and denominator are relatively prime, like $\frac{2}{3}$ or $\frac{17}{27}$, is in its simplest form. But a fraction like $\frac{6}{8}$ is not in simplest form, or lowest terms, because the numerator 6 and the denominator 8 have a common factor of 2.

$$\frac{6}{8} = \frac{3 \times 2}{4 \times 2}$$

The 2 in the numerator and the 2 in the denominator make a fraction that is equal to one whole. That 1, although disguised as $\frac{2}{2}$, is still worth 1, and multiplying by 1 doesn't change the value of a number.

$$\frac{6}{8} = \frac{3 \times 2}{4 \times 2} = \frac{3}{4} \times \frac{2}{2} = \frac{3}{4} \times 1 = \frac{3}{4}$$

To express a fraction in simplest form, or reduce it to its lowest terms, you can find the prime factorization of the numerator and the prime factorization of the denominator, and then eliminate the disguised ones.

$$\frac{120}{300} = \frac{2^3 \times 3 \times 5}{2^2 \times 3 \times 5^2} = \frac{2 \times 2 \times 2 \times 3 \times 5}{2 \times 2 \times 3 \times 5 \times 5} = \boxed{\frac{2}{2}} \times \boxed{\frac{2}{2}} \times \boxed{\frac{3}{3}} \times \frac{2}{5} \times \boxed{\frac{5}{5}} = 1 \times 1 \times 1 \times \frac{2}{5} \times 1 = \frac{2}{5}$$

For a faster method, find the greatest common factor of the numerator and denominator, and divide both the numerator and denominator by that GCF. The GCF for 120 and 300 is 60.

$$\frac{120}{300} = \frac{120 \div 60}{300 \div 60} = \frac{2}{5}$$

Sometimes you need to change the look of a fraction without changing its value, but you don't want to put it in simplest form. Instead, you want a larger denominator. You can change to an equivalent fraction with a different denominator by multiplying both the numerator and the denominator by the same number.

$$\frac{3}{7} = \frac{3 \times 5}{7 \times 5} = \frac{15}{35}$$
$$\frac{3}{7} = \frac{3 \times 13}{7 \times 13} = \frac{39}{91}$$

EXERCISE

3·1

Express each fraction in simplest form.

1. $\dfrac{45}{60}$

2. $\dfrac{27}{36}$

3. $\dfrac{42}{126}$

4. $\dfrac{28}{49}$

5. $\dfrac{75}{125}$

6. $\dfrac{189}{369}$

7. $\dfrac{88}{110}$

8. $\dfrac{35}{63}$

9. $\dfrac{105}{280}$

10. $\dfrac{153}{187}$

Change each fraction to an equivalent fraction with the indicated denominator.

11. $\dfrac{5}{6} = \dfrac{}{18}$

12. $\dfrac{4}{7} = \dfrac{}{35}$

13. $\dfrac{2}{9} = \dfrac{}{54}$

14. $\dfrac{3}{14} = \dfrac{}{84}$

15. $\dfrac{7}{10} = \dfrac{}{110}$

16. $\dfrac{1}{2} = \dfrac{}{24}$

17. $\dfrac{5}{16} = \dfrac{}{96}$

18. $\dfrac{7}{8} = \dfrac{}{48}$

19. $\dfrac{15}{16} = \dfrac{}{80}$

20. $\dfrac{11}{12} = \dfrac{}{132}$

Multiplying fractions

Multiplying by $\dfrac{1}{2}$, taking half of a number, has the same effect as dividing the number by 2. Half of 6 is 3, half of 20 is 10, and half of 500 is 250. Multiplying by a unit fraction like $\dfrac{1}{2}$ or $\dfrac{1}{3}$ is the equivalent of dividing by the denominator.

Half of $\dfrac{3}{4}$ is not as easy to picture. If you were asked for half of $\dfrac{6}{8}$, you could say that you had six things called eighths, so half of that would be three of those things, or $\dfrac{3}{8}$. But you don't want to say that half of $\dfrac{3}{4}$ is $1\dfrac{1}{2}$ quarters. That's just too many fractions. If you change $\dfrac{3}{4}$ to $\dfrac{6}{8}$, which is equivalent, you know half of $\dfrac{6}{8}$ is $\dfrac{3}{8}$.

$$\frac{1}{2} \times \frac{3}{4} = \frac{1}{2} \times \frac{3}{4} \times \frac{2}{2} = \frac{1}{2} \times \frac{6}{8} = \frac{3}{8}$$

To find $\dfrac{1}{7}$ of $\dfrac{3}{5}$ by this clunky method, you'd need to change $\dfrac{3}{5}$ to an equivalent form with a numerator divisible by 7.

$$\frac{1}{7} \times \frac{3}{5} = \frac{1}{7} \times \frac{3}{5} \times \frac{7}{7} = \frac{1}{7} \times \frac{21}{35} = \frac{3}{35}$$

Luckily, there's a shortcut, and you've probably already spotted it. $\dfrac{1}{2} \times \dfrac{3}{4} = \dfrac{1 \times 3}{2 \times 4} = \dfrac{3}{8}$ and $\dfrac{1}{7} \times \dfrac{3}{5} = \dfrac{1 \times 3}{7 \times 5} = \dfrac{3}{35}$.

To multiply fractions, multiply the numerators and multiply the denominators, then simplify the product, if possible.

To multiply a whole number and a fraction, write the whole number over a denominator of 1, and treat it as a fraction.

Perform each multiplication, and simplify the result if possible.

1. $\dfrac{6}{7} \times \dfrac{1}{3}$

2. $\dfrac{12}{17} \times \dfrac{1}{6}$

3. $\dfrac{15}{23} \times \dfrac{1}{5}$

4. $\dfrac{24}{35} \times \dfrac{1}{8}$

5. $\dfrac{9}{10} \times \dfrac{1}{3}$

6. $\dfrac{4}{5} \times \dfrac{1}{2}$

7. $\dfrac{14}{25} \times \dfrac{1}{7}$

8. $\dfrac{6}{7} \times \dfrac{1}{6}$

9. $\dfrac{39}{50} \times \dfrac{1}{13}$

10. $\dfrac{49}{100} \times \dfrac{1}{7}$

Multiplying by a fraction that is not a unit fraction follows the same rules. Multiplying by $\dfrac{5}{7}$ is equivalent to multiplying by $\dfrac{1}{7}$ and then multiplying by 5, so it's easy to put it all together.

$$\dfrac{5}{7} \times \dfrac{3}{5} = 5\left(\dfrac{1}{7} \times \dfrac{3}{5}\right) = 5\left(\dfrac{1 \times 3}{7 \times 5}\right) = \dfrac{5 \times 3}{7 \times 5} = \dfrac{15}{35}$$

When you multiply by a fraction that is not a unit fraction, there's a greater chance that you will need to simplify the result.

$$\dfrac{5}{7} \times \dfrac{3}{5} = \dfrac{5 \times 3}{7 \times 5} = \dfrac{15}{35} = \dfrac{3}{7}.$$

Perform each multiplication. Leave your answers in simplest form.

1. $\dfrac{3}{5} \times \dfrac{5}{7}$

2. $\dfrac{2}{9} \times \dfrac{3}{8}$

3. $\dfrac{7}{12} \times \dfrac{6}{7}$

4. $\dfrac{9}{11} \times \dfrac{2}{3}$

5. $\dfrac{2}{7} \times \dfrac{7}{12}$

6. $\dfrac{2}{3} \times \dfrac{15}{22}$

7. $\dfrac{3}{4} \times \dfrac{5}{21}$

8. $\dfrac{3}{5} \times \dfrac{10}{11}$

9. $\dfrac{9}{10} \times \dfrac{2}{3}$

10. $\dfrac{1}{10} \times \dfrac{5}{7}$

With a little bit of observation, you can get that simplifying done ahead of time. Once again, it's a matter of eliminating the disguised one. $\dfrac{5}{7} \times \dfrac{3}{5} = \dfrac{5 \times 3}{7 \times 5} = \dfrac{3}{7} \times \dfrac{5}{5} = \dfrac{3}{7} \times 1 = \dfrac{3}{7}$.

When you multiply fractions, you can cancel a factor from any numerator with a matching factor in any denominator.

$$\dfrac{14}{27} \times \dfrac{45}{56} = \dfrac{\cancel{2} \times \cancel{7}}{\cancel{3 \times 3} \times 3} \times \dfrac{\cancel{3 \times 3} \times 5}{\cancel{7} \times \cancel{2} \times 2 \times 2} = \dfrac{5}{3 \times 2 \times 2} = \dfrac{5}{12}$$

You'll get the same result if you multiply the numerators and multiply the denominators, and then simplify, but you'll have to do all the work of multiplying 14×45, and multiplying 27×56, and then figure out how to simplify $\dfrac{630}{1,512}$. It's a much easier job if you cancel first.

Perform each multiplication and express the answer in simplest form.

1. $\dfrac{3}{7} \times \dfrac{5}{8}$

2. $\dfrac{1}{4} \times \dfrac{5}{7}$

3. $\dfrac{2}{3} \times \dfrac{5}{8}$

4. $\dfrac{1}{5} \times \dfrac{10}{11}$

5. $\dfrac{12}{13} \times \dfrac{3}{4}$

6. $\dfrac{5}{8} \times \dfrac{16}{25}$

7. $\dfrac{7}{10} \times \dfrac{5}{14}$

9. $\dfrac{6}{7} \times \dfrac{49}{50}$

8. $\dfrac{8}{11} \times \dfrac{33}{64}$

10. $\dfrac{24}{25} \times \dfrac{35}{36}$

Dividing fractions

If multiplying by a unit fraction is the equivalent of dividing by the denominator, then dividing by a unit fraction is the equivalent of multiplying by the denominator. Dividing by a half is the same as multiplying by 2 and dividing by one-tenth is equivalent to multiplying by 10. There are two halves in one whole, and 15 thirds in five wholes.

$$24 \times \frac{1}{2} = 24 \div 2 = 12 \text{ and } 24 \div \frac{1}{2} = 24 \times 2 = 48$$

The rule for dividing by a fraction turns the problem on its head and makes it a multiplication problem, but it requires a definition first. If the product of two numbers is 1, each number is the **reciprocal** of the other. The reciprocal of 7 is $\dfrac{1}{7}$ because $7 \times \dfrac{1}{7} = 1$. The reciprocal of $\dfrac{2}{3}$ is $\dfrac{3}{2}$ because $\dfrac{2}{3} \times \dfrac{3}{2} = 1$. The reciprocal of a fraction can be found by switching the places of the numerator and denominator, or **inverting** the fraction.

To divide by a fraction, multiply by its reciprocal. Put another way, **to divide by a fraction, invert the divisor and multiply.**

To divide 12 by $\dfrac{2}{3}$, multiply 12 by $\dfrac{3}{2}$. (Put the whole number over a denominator of 1 to make it look like a fraction.)

$$12 \div \frac{2}{3} = \frac{12}{1} \times \frac{3}{2} = \frac{\overset{6}{\cancel{12}}}{1} \times \frac{3}{\underset{1}{\cancel{2}}} = \frac{18}{1} = 18$$

To divide $\dfrac{15}{49}$ by $\dfrac{5}{7}$, invert the divisor $\dfrac{5}{7}$ to get $\dfrac{7}{5}$, and multiply. Notice that the $\dfrac{15}{49}$, the dividend, doesn't change. As you did in earlier multiplication exercises, you can often simplify before multiplying, but you can multiply and then simplify if you prefer.

$$\frac{15}{49} \div \frac{5}{7} = \frac{15}{49} \times \frac{7}{5} = \frac{\overset{3}{\cancel{15}}}{\underset{7}{\cancel{49}}} \times \frac{\overset{1}{\cancel{7}}}{\underset{1}{\cancel{5}}} = \frac{3}{7}$$

EXERCISE
3·5

Perform each division and express the answer in simplest form.

1. $\dfrac{3}{5} \div \dfrac{1}{2}$

2. $\dfrac{5}{7} \div \dfrac{1}{5}$

3. $\dfrac{9}{10} \div \dfrac{3}{5}$

4. $\dfrac{15}{28} \div \dfrac{35}{48}$

5. $\dfrac{6}{7} \div \dfrac{3}{14}$

6. $\dfrac{8}{33} \div \dfrac{4}{11}$

7. $\dfrac{5}{21} \div \dfrac{11}{14}$

8. $\dfrac{8}{35} \div \dfrac{4}{15}$

9. $\dfrac{2}{7} \div \dfrac{4}{5}$

10. $\dfrac{25}{52} \div \dfrac{20}{39}$

Adding and subtracting fractions with common denominators

Adding and subtracting fractions is a simple matter when the denominators are the same. The denominator tells you what kind of things you have, and the numerator tells you how many of them you have. If the denominators are the same, you have all the same kind of things, and all you have to do is add or subtract the numerators, and keep the denominators. Simplify the result, if possible.

$$\frac{5}{8} + \frac{1}{8} = \frac{6}{8} = \frac{3}{4}$$
$$\frac{5}{8} - \frac{1}{8} = \frac{4}{8} = \frac{1}{2}$$

If the sum is an improper fraction, you can change to a mixed number.

$$\frac{6}{7} + \frac{3}{7} = \frac{9}{7} = 1\frac{2}{7}$$

EXERCISE
3·6

Add or subtract and express the answer in simplest form.

1. $\dfrac{4}{7} + \dfrac{1}{7}$

2. $\dfrac{6}{11} - \dfrac{4}{11}$

3. $\dfrac{4}{5} + \dfrac{2}{5}$

4. $\dfrac{18}{25} - \dfrac{3}{25}$

5. $\dfrac{6}{11} + \dfrac{3}{11}$

6. $\dfrac{7}{8} - \dfrac{3}{8}$

7. $\dfrac{7}{15} + \dfrac{4}{15}$

8. $\dfrac{19}{24} - \dfrac{11}{24}$

9. $\dfrac{34}{45} + \dfrac{11}{45}$

10. $\dfrac{17}{20} - \dfrac{9}{20}$

Adding and subtracting with different denominators

If you add six apples to a bowl containing eight apples, the bowl will have 14 apples. If you add six oranges to a bowl containing eight apples, it doesn't have 14 apples, or 14 oranges, or 14 appleoranges. It has 14 pieces of fruit, and fruit is the best you can do for a category in which they all fit. If you add $\frac{3}{8}$ to $\frac{2}{3}$, you won't get five eights or five thirds. In order to describe what you get, you've got to find a category in which both eighths and thirds fit. That category is called the **common denominator.**

The smallest common denominator for two fractions is the least common multiple of the two denominators. For $\frac{3}{8}$ and $\frac{2}{3}$, the least common denominator is 24. You don't have to find the smallest common denominator when you're adding or subtracting fractions, but if you use a bigger one, you'll have more simplifying to do at the end. If you used a common denominator of 48 or 240 to add $\frac{3}{8}$ and $\frac{2}{3}$, you'd have to work at putting the answer in simplest form. So, whenever possible, use the least common denominator to save work.

To add $\frac{3}{8}$ and $\frac{2}{3}$, you'll need to change the look of each fraction so that each fraction has a denominator of 24. $\frac{3}{8}=\frac{3}{8}\times\frac{3}{3}=\frac{9}{24}$ and $\frac{2}{3}=\frac{2}{3}\times\frac{8}{8}=\frac{16}{24}$. Once both fractions have the same denominator, add the numerators, and simplify if necessary.

$$\frac{3}{8}+\frac{2}{3}=\left(\frac{3}{8}\times\frac{3}{3}\right)+\left(\frac{2}{3}\times\frac{8}{8}\right)=\frac{9}{24}+\frac{16}{24}=\frac{25}{24}=1\frac{1}{24}$$

EXERCISE

3·7

Add or subtract and express the answer in simplest form.

1. $\frac{3}{4}+\frac{2}{5}$

2. $\frac{7}{8}-\frac{1}{3}$

3. $\frac{5}{12}+\frac{3}{4}$

4. $\frac{14}{15}-\frac{7}{9}$

5. $\frac{3}{7}+\frac{5}{21}$

6. $\frac{17}{21}-\frac{2}{3}$

7. $\frac{2}{3}+\frac{1}{4}$

8. $\frac{6}{7}-\frac{5}{14}$

9. $\frac{5}{18}+\frac{1}{6}$

10. $\frac{3}{28}-\frac{3}{35}$

Mixed numbers and improper fractions

Eight eighths make a whole, and six sixths make a whole. Eight sixths make more than a whole. There is a whole, and there are two sixths left over. Written as $\dfrac{8}{6}$, it's called an **improper fraction**, a fraction that is greater than one. Written as $1\dfrac{2}{6}$, it's a **mixed number**, a number made up of a whole number and a fraction.

To switch forms, you can think of the whole as eight eighths or six sixths, or whatever parts you're working with. $1\dfrac{2}{6} = \dfrac{6}{6} + \dfrac{2}{6} = \dfrac{8}{6}$. For a simple shortcut **to change a mixed number to an improper fraction, multiply the denominator times the whole number, and then add that to the numerator.**

$$1\dfrac{2}{6} = \dfrac{(6 \times 1) + 2}{6} = \dfrac{8}{6}$$

To change an improper fraction to a mixed number, divide the numerator by the denominator. The quotient will be the whole number part, and the remainder over the divisor will be the fraction part.

$$\dfrac{17}{5} \Rightarrow 17 \div 5 = 3 \text{ (with a remainder of 2)} \Rightarrow 3\dfrac{2}{5}$$

EXERCISE 3·8

Convert each improper fraction to a mixed number in its simplest form.

1. $\dfrac{64}{5}$

2. $\dfrac{19}{3}$

3. $\dfrac{46}{8}$

4. $\dfrac{140}{12}$

5. $\dfrac{14}{6}$

Convert each mixed number to an improper fraction.

6. $1\dfrac{3}{8}$

7. $5\dfrac{1}{2}$

8. $3\dfrac{5}{7}$

9. $2\dfrac{3}{5}$

10. $9\dfrac{1}{3}$

Multiplying and dividing mixed numbers

To multiply mixed numbers, first convert them to improper fractions. Multiply, following the rules for fractions, and if the result is an improper fraction, convert back to a mixed number.

$$2\frac{4}{5}\times7\frac{1}{2}=\frac{\overset{7}{\cancel{14}}}{\cancel{5}_{1}}\times\frac{\overset{3}{\cancel{15}}}{\cancel{2}_{1}}=\frac{21}{1}=21$$

To divide mixed numbers, change to improper fractions, follow the rules for dividing fractions, and if the result is an improper fraction, convert back to a mixed number.

$$6\frac{3}{8}\div4\frac{1}{4}=\frac{51}{8}\div\frac{17}{4}=\frac{\overset{3}{\cancel{51}}}{_{2}\cancel{8}}\times\frac{\cancel{4}^{1}}{\cancel{17}_{1}}=\frac{3}{2}=1\frac{1}{2}$$

EXERCISE 3·9

Multiply or divide as indicated. Leave answers in their simplest form.

1. $1\frac{1}{3}\times\frac{2}{5}$

2. $3\frac{1}{2}\times\frac{2}{3}$

3. $1\frac{3}{4}\times1\frac{1}{3}$

4. $5\frac{3}{8}\times2\frac{2}{3}$

5. $7\frac{1}{2}\times3\frac{3}{5}$

6. $2\frac{1}{3}\div\frac{1}{6}$

7. $4\frac{1}{5}\div\frac{7}{10}$

8. $\frac{5}{7}\div3\frac{3}{14}$

9. $6\frac{3}{7}\div1\frac{4}{5}$

10. $12\frac{2}{3}\div3\frac{1}{6}$

Adding and subtracting mixed numbers

To add mixed numbers in which the fraction parts have the same denominator, add the whole numbers, and add the fractions. Change any improper fractions to mixed numbers and simplify the results.

$$4\frac{1}{5}+3\frac{2}{5}=(4+3)+\left(\frac{1}{5}+\frac{2}{5}\right)=7+\frac{3}{5}=7\frac{3}{5}$$

$$1\frac{2}{3}+4\frac{2}{3}=(1+4)+\left(\frac{2}{3}+\frac{2}{3}\right)=5+\frac{4}{3}=5+1\frac{1}{3}=6\frac{1}{3}$$

Subtracting mixed numbers is sometimes as simple as subtracting the whole numbers and subtracting the fractions.

$$8\frac{7}{9} - 2\frac{5}{9} = (8-2) + \left(\frac{7}{9} - \frac{5}{9}\right) = 6\frac{2}{9}.$$

There are times, however, when you run into a problem. If you try to subtract $3\frac{1}{8} - 1\frac{7}{8}$ by subtracting the whole numbers and subtracting the fractions, you find yourself trying to subtract $\frac{7}{8}$ from $\frac{1}{8}$, which means subtracting 7 from 1, a larger number from a smaller one. You could try using negative numbers, but that seems to make the task more complicated.

There are two strategies you can use to get around the problem:

- **Improper fraction method** You can change the mixed numbers to improper fractions, subtract, and convert back to a mixed number. Simplify the result as needed.

$$3\frac{1}{8} - 1\frac{7}{8} = \frac{25}{8} - \frac{15}{8} = \frac{10}{8} = 1\frac{2}{8} = 1\frac{1}{4}$$

- **Regrouping method** The other way around the problem is to regroup, similar to the way you "borrow" when subtracting whole numbers. You take one from the whole number part and group it with the fraction, making an improper fraction.

$$3\frac{1}{8} = (2+1) + \frac{1}{8} = 2 + \left(\frac{8}{8} + \frac{1}{8}\right) = 2\frac{9}{8}.$$

Then you subtract.

$$3\frac{1}{8} - 1\frac{7}{8} = 2\frac{9}{8} - 1\frac{7}{8} = (2-1) + \left(\frac{9}{8} - \frac{7}{8}\right) = 1\frac{2}{8} = 1\frac{1}{4}$$

EXERCISE

3·10

Add or subtract as indicated. Simplify your answers.

1. $1\frac{3}{5} + 2\frac{1}{5}$

2. $5\frac{12}{13} - 3\frac{2}{13}$

3. $8\frac{5}{6} + 5\frac{5}{6}$

4. $9\frac{1}{12} - 5\frac{5}{12}$

5. $5\frac{21}{32} + 4\frac{31}{32}$

6. $6\frac{3}{8} - 3\frac{5}{8}$

7. $12\frac{3}{7} + 9\frac{1}{14}$

8. $5\frac{5}{6} - 1\frac{2}{3}$

9. $11\dfrac{3}{4} + 8\dfrac{5}{12}$

10. $8\dfrac{3}{11} - 4\dfrac{2}{5}$

11. $4\dfrac{9}{10} - 1\dfrac{2}{7}$

12. $5\dfrac{1}{6} + 3\dfrac{2}{5}$

13. $14\dfrac{9}{11} - 6\dfrac{2}{3}$

14. $\dfrac{19}{28} + 7\dfrac{3}{7}$

15. $11\dfrac{1}{3} - 4\dfrac{7}{8}$

Decimals

Our number system is a decimal, or base ten, system. We start with 1 (or 0) and count up through 2, 3, 4, 5, 6, 7, 8, and 9, but after that, we need another digit, and another place to put it. By writing 10, which we read as "ten," we're saying we have 1 group of ten and 0 ones. Groups of ten are key to our system.

Place value

Two digits let us write numbers from 10 to 99, 9 groups of ten and 9 ones. To go higher we need another place, worth ten tens or one hundred. To go beyond 999 we need a place for 10×100 or 1,000. Each new slot for a digit, or decimal place, that we introduce has a value 10 times that of the place to its right.

$$\overline{1,000,000} \quad \overline{100,000} \quad \overline{10,000} \quad \overline{1,000} \quad \overline{100} \quad \overline{10} \quad \overline{1}$$

Putting digits into these places allows us to express a number compactly. We don't have to say 4 groups of 100 and 2 groups of ten and 7 ones. We just write 427, and the position of the digits tell us what they are worth. The number 427 is different from 274 or 742. In 427, the 7 is 7 ones. In 274, the 7 is 7 tens and in 742, it's 7 hundreds. In the number below, read "three million, seven hundred forty-one thousand, five hundred ninety-six," the 7 tells us there are 7 groups of 100,000 each, the 5 says there are 5 groups of 100 and the 6 is worth 6 ones.

$$3,741,596 = \underset{1,000,000}{\underline{3}} \quad \underset{100,000}{\underline{7}} \quad \underset{10,000}{\underline{4}} \quad \underset{1,000}{\underline{1}} \quad \underset{100}{\underline{5}} \quad \underset{10}{\underline{9}} \quad \underset{1}{\underline{6}}$$

Decimal fractions

What we commonly call decimals are decimal fractions, a translation of fractions into our decimal system. Just as each place to the left of the decimal point represents a power of ten, each place to the right of the decimal point represents a fraction with a denominator that is a power of ten. The first digit after the decimal point tells the number of tenths, the second the number of hundredths, then thousandths, ten-thousandths, and so on.

$$\overline{1,000,000} \quad \overline{100,000} \quad \overline{10,000} \quad \overline{1,000} \quad \overline{100} \quad \overline{10} \quad \overline{1} \, \cdot \, \overline{\tfrac{1}{10}} \quad \overline{\tfrac{1}{100}} \quad \overline{\tfrac{1}{1,000}} \quad \overline{\tfrac{1}{10,000}} \quad \overline{\tfrac{1}{100,000}}$$

The number 7.921 represents 7 ones, 9 tenths $\left(\dfrac{9}{10}\right)$, 2 hundredths $\left(\dfrac{2}{100}\right)$, and 1 thousandth $\left(\dfrac{1}{1,000}\right)$. If you add the fractions $\dfrac{9}{10} + \dfrac{2}{100} + \dfrac{1}{1,000} = \dfrac{900}{1,000} + \dfrac{20}{1,000} + \dfrac{1}{1,000} = \dfrac{921}{1,000}$, you can see that the decimal fraction is nine hundred twenty-one thousandths.

$$7.921 = 7.\ \ \underset{1}{9}\ \ \underset{\big/10}{2}\ \ \underset{\big/100}{1}\ \ \underset{\big/1,000}{} = 7 + \frac{9}{10} + \frac{2}{100} + \frac{1}{1,000}$$

The number 6.45 is 6 ones and 4 tenths and 5 hundredths, or 6 and 45 hundredths. The number 0.0007 is 7 ten-thousandths.

The place in which the decimal fraction ends tells you its "denominator."

4.831 is 4 and 831 *thousandths*. 0.4831 is 4,831 *ten-thousandths*. 483.1 is four hundred eighty-three and 1 *tenth*. 0.00004831 is 4,831 *hundred-millionths*.

The decimal point between the ones place and the tenths place is read as "and."

5.73 is five *and* seventy-three hundredths. 1.000001 is 1 *and* 1 millionth.

EXERCISE 4·1

Write in words the name of each number as you would read it.

1. 4.37

2. 16.025

3. 2.009

4. 8.7102

5. 1.426

Write each number. Make sure each digit is in the correct decimal place.

6. Eight and thirty-seven hundredths

7. Four and twelve ten-thousandths

8. Five and two hundred-thousandths

9. Three hundred and forty-five thousandths

10. Six and three hundred forty-five thousandths

Fraction to decimal

To change a fraction to its decimal fraction equivalent, divide the numerator by the denominator. Remember that every fraction is a statement of division. You change a fraction to a decimal by doing that division.

Terminating decimals

To change $\frac{7}{8}$ to a decimal, divide 7 by 8. Set up the division, with a divisor of 8 and a dividend of 7. Place the decimal point after the 7 and add zeros. $8\overline{)7.000}$. (If you're not sure how many zeros to add, just start with a few. You can add more later if you need them or drop extras if you don't need them.) Let the decimal point move straight up from the dividend to the quotient. Start dividing and keep dividing until you get a zero remainder or you start to see a pattern repeating.

$$
\begin{array}{r}
0.875 \\
8\overline{)7.000} \\
\underline{64} \\
60 \\
\underline{56} \\
40 \\
\underline{40} \\
0
\end{array}
$$

EXERCISE 4·2

Use long division to convert each fraction to a terminating decimal.

1. $\frac{3}{4}$

2. $\frac{2}{5}$

3. $\frac{16}{25}$

4. $\frac{3}{20}$

5. $\frac{3}{16}$

6. $\frac{7}{100}$

7. $\frac{3}{8}$

8. $\frac{7}{32}$

9. $\frac{1}{80}$

10. $\frac{11}{200}$

Repeating decimals

Not every attempt to change a fraction to a decimal will terminate with a zero remainder. Some fractions convert to what we call repeating decimals. The fraction $\frac{1}{3}$, for example, when you divide 1 by 3, will give you 0.333...., and no matter how many zeros you add and how long you keep dividing, you'll never get a zero remainder. The decimal equivalent of $\frac{1}{3}$ is an endless string of threes behind the decimal point.

Repeating decimals won't end, but they will repeat a pattern of digits. It may be a single digit that repeats, as for $\frac{1}{3}$, or it may be a group of digits, but any number that can be expressed as a fraction will, when changed to a decimal, either terminate or repeat, so keep dividing until you see the pattern.

A number that can be written as the quotient of two integers, that is, as a fraction, is called a rational number.
Every rational number, if changed to a decimal, either terminates or repeats a pattern.

To change $\frac{5}{6}$ to a decimal, divide 5 by 6, adding a decimal point and zeros. Watch for the repeating digit.

$$
\begin{array}{r}
0.833 \\
6\overline{)5.000} \\
\underline{48} \\
20 \\
\underline{18} \\
20 \\
\underline{18} \\
2
\end{array}
$$

Once you see that the pattern is going to repeat, you can stop dividing and write the decimal using a bar to show the repeating digit or pattern. The fraction $\frac{1}{3}$ can be written as $0.\overline{3}$ and $\frac{5}{6} = 0.8\overline{3}$. The bar over the 3 tells us that the 3 repeats, and the fact that the bar is not over the 8 says that 8 is not part of the repeating pattern. When there is more than one digit in the pattern, the bar is placed over the entire pattern. $\frac{1}{7} = 0.\overline{142857}$.

EXERCISE 4·3

Change each fraction to the equivalent decimal. Use long division and keep dividing until you can see the pattern. Write the decimal using a bar to show the digits that repeat.

1. $\dfrac{2}{15}$

2. $\dfrac{5}{18}$

3. $\dfrac{8}{9}$

4. $\dfrac{2}{3}$

5. $\dfrac{4}{7}$

6. $\dfrac{5}{12}$

7. $\dfrac{13}{60}$

8. $\dfrac{4}{27}$

9. $\dfrac{1}{144}$

10. $\dfrac{7}{30}$

Decimal to fraction

To change a decimal that terminates to its equivalent fraction, listen to its proper name. When people see 0.87, they'll often say "point eighty-seven," but the proper name of that decimal is eighty-seven hundredths. That proper name tells you exactly how to write it as a fraction: $\dfrac{87}{100}$. The only other work you might have to do is simplifying. The decimal 0.125 is one hundred twenty-five thousandths, or $\dfrac{125}{1,000}$, but that fraction can be simplified. $\dfrac{125}{1,000} = \dfrac{1}{8}$.

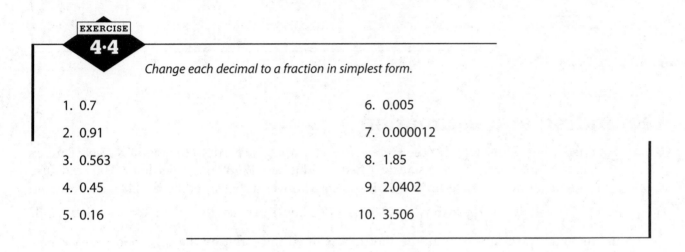

EXERCISE 4·4

Change each decimal to a fraction in simplest form.

1. 0.7

2. 0.91

3. 0.563

4. 0.45

5. 0.16

6. 0.005

7. 0.000012

8. 1.85

9. 2.0402

10. 3.506

Repeating decimals

The process for changing a repeating decimal to an equivalent fraction is a little more complicated, but only because there are variations. If every digit in the decimal is part of the repeating pattern, like $0.\overline{3}$ or $5.\overline{29}$, place the digits in the pattern in the numerator and the same number of nines in the denominator. Simplify if possible. $0.\overline{3} = \dfrac{3}{9} = \dfrac{1}{3}$ and $5.\overline{29} = 5\dfrac{29}{99}$.

If the pattern doesn't begin right after the decimal point, so that a few digits occur after the decimal but before the repeating pattern begins, you'll need to do a little calculating first. Let's look at the decimal $0.12\overline{3}$:

- Write the digits before the pattern and one pattern as a whole number (123).
- From that, subtract the number formed by the digit before the pattern (12).
  - Subtract $123 - 12 = 111$. That will be the numerator.
- For the denominator, use a 9 for each digit in the pattern, followed by a 0 for each digit before the pattern.
  - One digit (3) in the pattern and 2 digits (12) before the pattern, so the denominator is one 9 and two zeros.
  - $0.12\overline{3} = \dfrac{111}{900}$.

To convert $0.12\overline{345}$ to a decimal,

- Subtract $12,345 - 123 = 12,222$ for the numerator.
- Put that over a denominator of 99,000.
  - $0.12\overline{345} = \dfrac{12,222}{99,000} = \dfrac{679}{5,500}$.

Change each decimal to an equivalent fraction in simplest form.

1. $0.\overline{1}$

2. $0.\overline{18}$

3. $0.\overline{037}$

4. $0.2\overline{27}$

5. $0.\overline{285714}$

6. $0.0\overline{857142}$

7. $0.0\overline{1923076}$

8. $0.\overline{6}$

9. $0.\overline{54}$

10. $0.4\overline{6}$

Rounding and estimation

You might look up the fact that the average distance from the earth to the moon is 238,857 miles and that the speed of the space shuttle, when in orbit, was 17,285 miles per hour. You could get your calculator out and determine that, if you could travel at that speed, it would take an average of $13\dfrac{14,152}{17,285}$ hours or approximately 13.8187445762 hours to reach the moon. On the other hand, if you could travel at the speed the shuttle was lifted into orbit by the solid booster rockets – 3,094 miles per hour – the trip would take $77\dfrac{619}{3,094}$ hours or approximately 77.20006464 hours. (The second is the more reasonable, by the way.)

If those awkward fractions and long decimals left you feeling that 13 or 77 was as much information as you really needed, then like most people, you recognize that often an estimate is adequate, especially when you're dealing with very large or very small numbers. When you work with exponential expressions, the numbers will often get very large or very small, so estimating will be helpful. To arrive at a reasonable estimate, you'll want to round the numbers involved.

Rounding isn't just randomly saying "it's about this much," although even that can be helpful sometimes. Rounding is a system to place a number between two boundaries and determine which end of the interval the number is closer to. It starts with the decision to round to the nearest _____ and choosing a decimal place to fill the blank. For example, you might choose to round to the nearest million or the nearest ten-thousand. If you choose to round 13 to the nearest ten, you place the number in an interval from 10, the multiple of ten just below 13, to 20, the multiple of ten just above 13. The decision you're making is whether 13 is closer to 10 or closer to 20, and in the case of 13, it's closer to 10. The number 13, rounded to the nearest 10, is 10, because 13 is closer to 10 than to 20. On the other hand, 17 rounded to the nearest 10 is 20, because 17 is closer to 20 than to 10.

Of course, many situations in which you make a choice to round involve very large or very small numbers. You might round the average distance from the earth to the moon, which is 238,857 miles, to 200,000 miles. That says that you want to focus on the largest decimal place, the hundred-thousands, and that 238,857 is closer to 200,000 than to 300,000. You might round it to the nearest ten-thousand, by deciding if it's closer to 230,000 or 240,000, the "round" numbers below it or above it.

To round a number to a particular place:

- ◆ Determine the decimal place to which you want to round.
- ◆ Visualize the interval your number falls in; for example, between 200,000 and 300,000 or between 238,000 and 239,000.
- ◆ Look to the digit following the place to which you're rounding.
  - ■ If the digit is 5 or more, round up; that is, declare that your number is closer to the upper end of the interval.
  - ■ If it's 4 or less, round down by declaring the number closer to the lower end of the interval.

The number 238,857, if rounded to the nearest ten-thousand, rounds up to 240,000 because the 8 in the thousands place is greater than 5, telling you that 238,857 is closer to 240,000 than 230,000. The number 3,094, if rounded to the nearest thousand would be 3,000 because the digit following the thousands place is a 0, telling you to round down.

The process for rounding decimals is the same:

- ◆ Determine the decimal place to which you want to round. You might round 0.0034 to the nearest thousandth, or round 6.879 to the nearest tenth or nearest hundredth.
- ◆ Visualize the interval your number falls in; for example, between 0.003 and 0.004, or between 6.8 and 6.9. You can add trailing zeros, like 0.0030 or 6.800, if it makes it easier for you to understand the interval.
- ◆ Look to the digit following the place to which you're rounding.
  - ■ If the digit is 5 or more, round up; that is, declare that your number is closer to the upper end of the interval.
  - ■ If it's 4 or less, round down by declaring the number closer to the lower end of the interval.

To round 0.0034 to the nearest thousandth, look to the digit in the ten-thousandths place, which is 4. That's less than five, so round down to 0.003.

To round 6.879 to the nearest tenth, look at the digit in the hundredths place, which is 7, and round up because 7 is greater than 5. Rounding 6.879 to the nearest tenth gives you 6.900 or simply 6.9.

Note that the 9 in the thousandths place of 6.879 plays no role when you're rounding to the nearest tenth. You're placing 6.879 in the interval from 6.800 and 6.900. It doesn't matter if you have 6.879 or 6.871. Only the digit immediately after the place to which you're rounding matters.

Estimating calculations

The average distance from earth to the moon rounded to the nearest ten-thousand is 240,000 and the orbital speed of the shuttle, 17,285 miles per hour, rounded to the nearest ten-thousand is 20,000. With those estimates, you can easily divide 240,000 ÷ 20,000 and get an estimated travel time of 12 hours. That estimate is a bit less than the actual calculation, but close enough for most needs.

If you want to give everyone in your class a "stress ball" to help them through exams and you can buy the balls for $0.16 each. There are 38 students in your class. What will this idea cost? Round $0.16 to $0.20 and 38 to 40 and you can estimate the cost as $0.20 × 40 = $8.00. The actual cost would be $0.16 × 38 = $6.08, so you'll get change.

In exercises 1 through 5, fill in the first two blanks with numbers that surround the given number, and fill the last blank with the number to which you would round.

1. When rounding 47,839.164 to the nearest thousand, it is between _____ and _____, closer to _____.

2. When rounding 47,839.164 to the nearest hundredth, it is between _____ and _____, closer to _____.

3. When rounding 47,839.164 to the nearest ten-thousand, it is between _____ and _____, closer to _____.

4. When rounding 149.08403 to the nearest tenth, it is between _____ and _____, closer to _____.

5. When rounding 149.08403 to the nearest ten-thousandth, it is between _____ and _____, closer to _____.

Round each number to the indicated place.

6. 9.43598 to the nearest thousandth

7. 0.908319 to the nearest hundredth

8. 14,668.7 to the nearest thousand

9. 5,147.02 to the nearest hundred

10. 405,810 to the nearest ten-thousand

11. 0.733813 to the nearest ten-thousandth

12. 43,992 to the nearest hundred

13. 339,362 to the nearest thousand

14. 995.467 to the nearest tenth

15. 2.0034 to the nearest hundredth

Estimate the result of each calculation.

16. $37,542 \times 18$

17. 3.469×8.71

18. $63.475 \div 8.773$

19. $591,278,634 \div 214,865$

20. 0.01637×0.004205

Adding and subtracting

To add or subtract numbers involving decimals, position the numbers vertically, with the decimal points aligned. This will assure that the digits in the same decimal place are lined up. Add or subtract as you would for whole numbers and place the decimal point in the answer directly under those in the problem.

To add $72.392 + 8.57$, position the numbers one under the other, with the decimal points aligned. You can add zeros if the empty spaces bother you.

$$
\begin{array}{r}
72.392 \\
+\ \ 8.570 \\
\end{array}
$$

Add just as you would for whole numbers and bring the decimal point straight down.

$$
\begin{array}{r}
72.392 \\
+\ \ 8.570 \\
\hline
80.962 \\
\end{array}
$$

To subtract $26.34 - 18.225$, line up the numbers with the decimal points one over the other. When you're subtracting it's a good idea to add zeros to give all the numbers the same number of digits. Then subtract normally and bring the decimal point straight down.

$$
\begin{array}{r}
26.340 \\
-18.225 \\
\hline
8.115 \\
\end{array}
$$

EXERCISE 4·7

Add.

1. $4.39 + 3.571$

2. $19.3701 + 7.92$

3. $102.396 + 83.2$

Subtract:

6. $102.396 - 83.2$

7. $4.39 - 3.571$

8. $873.992 - 175.2779$

4. $16.009 + 47.0001$

5. $873.992 + 75.2779$

9. $19.3701 - 7.92$

10. $47.0001 - 16.009$

Multiplying

When you multiply $\frac{1}{10} \times \frac{1}{10}$, you get $\frac{1}{100}$. If you think about that same problem in decimal terms, it's $0.1 \times 0.1 = 0.01$. The decimals in the problem each had one digit after the decimal point, but the answer has two digits after the decimal point. If you multiply $\frac{1}{100} \times \frac{1}{10}$, or 0.01×0.1, you get $\frac{1}{1,000}$ or 0.001. The numbers in the problem have two digits and one digit after the decimal point, and there are three digits after the decimal point in the answer.

To multiply decimals, multiply the numbers as though no decimals were present. Then count the number of digits after the decimal point in each factor and add those together. Place the decimal point in the product so that the number of digits after the decimal point is the total of the number of decimal digits in the problem.

To multiply 459.72 × 8.37, begin by multiplying as though you were multiplying 45,972 by 837. Don't worry about decimal points yet. Then, since each factor has two digits after the decimal point, and 2 + 2 = 4, place the decimal point so that there are four digits after the decimal point in the product.

$$
\begin{array}{r}
459.72 \\
\times\ 8.37 \\
\hline
321804 \\
1379160 \\
\underline{36777600} \\
3847.8564
\end{array}
$$

It's always helpful to make an estimate of the product to help you place the decimal point. In the example above, 459.72 is between 400 and 500, and 8.37 is a bit more than 8, so the product will be between 400 × 8 and 500 × 8, or between 3,200 and 4,000. When you're faced with the digits 38478564, and you're trying to place the decimal point, knowing that your product should be between 3,200 and 4,000 will help you place the decimal point as 3,847.8564.

If the problem had been 4597.2 × 83.7 (rather than 459.72 × 8.37), you would have estimated it as about 5,000 × 80 or about 400,000 and would want to have two digits after the decimal point in the answer. You'll place the decimal point between the 5 and the 6 in 38478564, getting a product of 384,785.64.

EXERCISE

4·8

Estimate the product by rounding and multiplying.

1. 3.2 × 4.8

2. 12.3 × 29.8

3. 84.29 × 16.3

4. 423.8 × 2.7

5. 415.81 × 24.6

Multiply.

6. 5.5 × 8.6

7. 9.8 × 0.28

8. 0.27 × 0.012

9. 0.072 × 4.2

10. 3.001 × 9.08

11. 8.4 × 8.93

12. 0.62 × 21.1

13. 0.099 × 0.505

14. 7.003 × 0.0357

15. 9.03 × 0.80306

If 238 × 45 = 10710, find each of the following products.

16. 2.38 × 4.5

17. 2.38 × 0.45

18. 23.8 × 0.45

19. 0.238 × 4.5

20. 0.0238 × 0.045

Dividing

The rule for dividing by a decimal is simple: you don't. You can divide a decimal by a whole number. You just divide as though the decimal point were not there, and then place the decimal point in the quotient directly above the decimal point in the dividend. It's a good idea to estimate the quotient before you divide, to help you place the decimal point correctly.

Dividing a decimal by a whole number

To divide 592.76 by 14, first make an estimate. 592.76 is close to 600, and 14 is close to 15. 600 ÷ 15 = 40, so expect a quotient around 40. Then set up your division and divide without worrying about the decimal point. Finally, move the decimal point straight up.

$$
\begin{array}{r}
42.34 \\
\uparrow \\
14\overline{)592.76} \\
\underline{56} \\
32 \\
\underline{28} \\
47 \\
\underline{42} \\
56 \\
\underline{56}
\end{array}
$$

Dividing by a decimal

But there is no rule for dividing a number by a decimal. To do a division problem that has a decimal divisor, you first transform it to a problem with the same quotient and a whole number divisor. It may sound like you need to find your magic wand, but it's all about those disguised ones we used when we changed the denominator of a fraction.

To divide 4,821.6 by 7.35, first estimate the quotient. If you round 4,821.6 to 4,900 and 7.35 to 7, you'll have 4,900 ÷ 7 = 700. Expect a quotient around 700. Then think about the division problem written as a fraction: $\dfrac{4{,}821.6}{7.35}$. Multiplying the numerator and the denominator by the same number will change the look of the fraction without changing its value. Multiplying each of these numbers by 100 will turn them into whole numbers. $\dfrac{4{,}821.6}{7.35} \times \dfrac{100}{100} = \dfrac{482{,}160}{735}$. Work through that whole number division.

$$
\begin{array}{r}
656 \\
735\overline{)482160} \\
\underline{4410} \\
4116 \\
\underline{3675} \\
4410 \\
\underline{4410}
\end{array}
$$

Remember your estimate of 700 and remember that multiplying the dividend and the divisor by 100 moved the decimal point to the end of each number. You can let the decimal point float straight up and you'll find that $4{,}821.6 \div 7.35 = 656$.

You don't actually have to write your division problems as fractions every time you have a decimal divisor. Just use this shortcut.

- To divide by a decimal, move the decimal point to the end of the divisor.
- Count the number of places you've moved the decimal point and move the same number of places in the dividend, adding zeros if necessary.
- Divide and bring the decimal point straight up.

To divide 586.32 by 4.2, first estimate. $500 \div 4 \approx 125$ and $600 \div 4 \approx 150$, so expect a quotient between 125 and 150. Move the decimal point in 4.2 one place right to make it 42. You've actually multiplied by 10. Move the decimal in 586.32 one place to the right, making it 5,863.2, so that you're multiplying both numbers by 10. Then divide 5,863.2 by 42.

$$
\begin{array}{r}
139.6 \\
42\overline{)5863.2} \\
\underline{42} \\
166 \\
\underline{126} \\
403 \\
\underline{378} \\
252 \\
\underline{252} \\
0
\end{array}
$$

If you needed to divide 5863.2 by 4.2, your result would be 1,396, and if you wanted to divide 58.632 by 42, you'd get 1.396.

Estimate each quotient.

1. $24.57 \div 3$
2. $18.69 \div 12$
3. $71.8 \div 8.9$

4. $0.57 \div 2.9$
5. $66.3 \div 8.5$

Perform each division.

6. $93.3 \div 0.24$
7. $5.181176 \div 1.001$
8. $7.7959 \div 3.01$

9. $8.0206 \div 0.063$
10. $6.65 \div 0.0099$

If 1,575 ÷ 63 = 25, find each of the following quotients.

11. $157.5 \div 63$
12. $15.75 \div 6.3$
13. $157.5 \div 0.63$

14. $15.75 \div 0.063$
15. $1.575 \div 0.00063$

Quick tricks

Because our decimal system is based on tens, multiplying or dividing by 10 or a power of 10 is a special case. When you multiply 3.792×10, you get 37.92, the same digits with the decimal point moved one place right.

To multiply a number by 10, move the decimal point one place to the right, adding a zero if necessary.

$$6.8 \times 10 = 68 \qquad 74 \times 10 = 740.$$

To multiply by a power of 10, move the decimal point one place right for each power or each zero in the multiplier.

♦ To multiply by 100, move the decimal point two places.
♦ To multiply by 1,000, move the decimal point three places to the right.

$$3.972 \times 100 = 397.2 \qquad 1.06 \times 1,000 = 1060.$$

To divide by 10 or a power of ten, move the decimal point to the left, one place for 10, two places for 100 and so on. Add leading zeros as needed.

$$42.78 \div 10 = 4.278 \qquad 53 \div 100 = 0.53 \qquad 7.1 \div 1,000 = 0.0071$$

Multiply or divide by moving the decimal point.

1. 9.4×100

2. $82.7 \times 10{,}000$

3. 16.32×10

4. $118.902 \times 1{,}000$

5. $414.87 \times 100{,}000$

6. $119.42 \div 10$

7. $185.001 \div 100$

8. $1.92 \div 1{,}000$

9. $0.603 \div 10{,}000$

10. $1213.49 \div 100{,}000$

Ratios, proportions, and percents

One reason to write a number like $\frac{3}{5}$ is to talk about a fraction, a part of some whole. If a starting pitcher pitches 6 innings of a 10-inning game, he pitches $\frac{6}{10}$ or $\frac{3}{5}$ of the game. But the same notation can also be used to compare numbers. If a chorus is made up of 6 men and 10 women, you can say that the men represent $\frac{6}{16}$ or $\frac{3}{8}$ of the chorus—a part—but you can also compare the number of men to the number of women by saying that the ratio of men to women is "6 to 10." You can write that as 6:10 or $\frac{6}{10}$, and you can simplify either one to 3:5 or $\frac{3}{5}$. A ratio is a comparison of two numbers by division, often written like a fraction.

Ratio and proportion

Suppose you know that for your favorite summer drink, the ratio of sparkling water to fruit juice is 1:4. That says you're using one ounce of sparkling water for every four ounces of juice. But if you're throwing a party and want to make a big batch all at once, instead of glass by glass, you need to translate that 1:4 into more useful terms. You can do that just the way you change the look of a fraction. Suppose you have 64 ounces of sparkling water. How much juice will you need?

Think of the ratio as a fraction, $\frac{1}{4}$, and realize that you want to change the look of that fraction so that the numerator is 64. $\frac{1}{4} \times \frac{64}{64} = \frac{64}{256}$. For 64 ounces of sparkling water, you'll want 256 ounces of juice. You know that the drink will taste the same because the ratio $\frac{1}{4}$ is equal to the ratio $\frac{64}{256}$. You've made a bigger batch of the drink, but you've kept the same proportions. You've adjusted the amounts proportionally.

A **proportion** is two equal ratios. If you write your ratios like fractions, a proportion looks like two equal fractions. $\frac{1}{4} = \frac{64}{256}$. If you write the ratios using colons, you write 1:4 = 64:256. (In some older books, you'll see 1:4 :: 64:256, but that double colon notation isn't used much anymore.)

When a proportion is written with colons, for example, 2:5 = 14:35, the first and last numbers, the far ends, are called the **extremes**. The 2 and the 35 are the extremes.

The two middle numbers, the 5 and the 14, are called the **means**. When the proportion is written as two equal fractions, for example, $\frac{2}{5} = \frac{14}{35}$, you have $\frac{\text{extreme}}{\text{mean}} = \frac{\text{mean}}{\text{extreme}}$.

Those labels, *means* and *extremes*, just make it easier to explain the most important property of proportions, which is officially called the **means-extremes property**: in any proportion, the product of the means equals the product of the extremes. In the proportion $\frac{2}{5} = \frac{14}{35}$, the product of the means is $5 \times 14 = 70$ and the product of the extremes is $2 \times 35 = 70$. In the proportion $\frac{1}{4} = \frac{64}{256}$, $4 \times 64 = 256$ and $1 \times 256 = 256$.

If you remember that a proportion is two equal ratios, or two equal fractions, it's not hard to see why that's true. Look at $\frac{1}{4} = \frac{64}{256}$, but remember that we got that second fraction by multiplying the first one by $\frac{64}{64}$, so rewrite it this way: $\frac{1}{4} = \frac{1 \times 64}{4 \times 64}$. The product of the means is $4 \times 1 \times 64$ and the product of the extremes is $1 \times 4 \times 64$. The proportion is

$$\frac{\text{numerator}}{\text{denominator}} = \frac{\text{numerator} \times \text{multiplier}}{\text{denominator} \times \text{multiplier}}$$

The product of the means is denominator $\times$ numerator $\times$ multiplier, and the product of the extremes is numerator $\times$ denominator $\times$ multiplier.

You can test whether you have a proportion by checking to see if the fractions are equal. Do they simplify to the same fraction? Or you can test by checking to see if the product of the means equals the product of the extremes.

EXERCISE

5·1

For questions 1 through 5, decide whether each statement is a proportion by simplifying both fractions.

1. $\dfrac{2}{3} = \dfrac{24}{36}$

2. $\dfrac{11}{15} = \dfrac{22}{33}$

3. $\dfrac{8}{15} = \dfrac{72}{135}$

4. $\dfrac{24}{75} = \dfrac{120}{375}$

5. $\dfrac{143}{169} = \dfrac{132}{156}$

For questions 6 through 10, decide whether each statement is a proportion by finding the product of the means and the product of the extremes.

6. $\dfrac{12}{21} = \dfrac{8}{14}$

7. $\dfrac{6}{8} = \dfrac{72}{98}$

8. $\dfrac{3}{11} = \dfrac{18}{65}$

9. $\dfrac{120}{150} = \dfrac{108}{135}$

10. $\dfrac{4}{123} = \dfrac{12}{41}$

Properties of proportions

The means-extremes property is the most important property of proportions, but there are others that can be useful.

The **flip property** says that if you have a proportion $\frac{a}{b} = \frac{c}{d}$, then you can flip or invert both fractions and you'll still have a proportion. $\frac{b}{a} = \frac{d}{c}$

The **swap property** says that if you have a proportion $\frac{a}{b} = \frac{c}{d}$, you can swap the means $\frac{a}{c} = \frac{b}{d}$ or swap the extremes $\frac{d}{b} = \frac{c}{a}$ and you'll still have a proportion.

The **addition (or subtraction) property** says that if you have a proportion $\frac{a}{b} = \frac{c}{d}$, you can add the denominator of each ratio to the numerator $\frac{a \pm b}{b} = \frac{c \pm d}{d}$, and you'll still have a proportion. The same is true for subtracting the denominator from the numerator. And it will also work if you add the numerator to the denominator. So if $\frac{a}{b} = \frac{c}{d}$ is a proportion, so is $\frac{a}{b \pm a} = \frac{c}{d \pm c}$.

Of course, since a proportion is two equal ratios, changing the look of one or both ratios without changing value will create a new proportion. Call it the equal ratios property. If $\frac{a}{b} = \frac{c}{d}$ is a proportion, then $\frac{ax}{bx} = \frac{cy}{dy}$ is a proportion.

EXERCISE
5·2

The statement $\frac{8}{15} = \frac{72}{135}$ is a proportion. For each of the following statements, decide if the statement is a proportion, and if it is, tell which property has been applied.

1. $\frac{8}{72} = \frac{15}{135}$

2. $\frac{15}{8} = \frac{135}{72}$

3. $\frac{23}{15} = \frac{87}{135}$

4. $\frac{8}{7} = \frac{72}{63}$

5. $\frac{23}{15} = \frac{207}{135}$

6. $\frac{135}{15} = \frac{72}{8}$

7. $\frac{72}{15} = \frac{8}{135}$

8. $\frac{8}{15} = \frac{64}{120}$

9. $\frac{8}{23} = \frac{72}{143}$

10. $\frac{15}{8} = \frac{72}{135}$

Finding the missing piece

When you're working with proportions, the most common question you want to answer is not "is this a proportion?" but "what number completes this proportion?" You saw that earlier, thinking about the amount of juice needed to mix with 64 ounces of sparkling water. In that example, it was easy to find an equivalent fraction. The ratio of water to juice was 1:4, or $\frac{1}{4}$, and you could just multiply to find the equal fraction. $\frac{1}{4} \times \frac{64}{64} = \frac{64}{256}$. The question was what number completes the proportion $\frac{1}{4} = \frac{64}{?}$, and by multiplying, you could find that the missing number was 256. That's one way to find a missing number in a proportion.

Often, the numbers are not so cooperative. Suppose you're mixing paint, and you know that to get a particular shade of orange you need to mix yellow and red in a ratio of $\frac{5}{7}$. You need a large quantity of the paint, and you have a quart, or 32 ounces, of the red paint. How much yellow do you need? You could figure out what to multiply by changing the denominator of 7 to a denominator of 32, but that's not going to be pretty.

The easier way to determine what number completes the proportion $\frac{5}{7} = \frac{?}{32}$ is to use the means-extremes property, or what's commonly called **cross-multiplying**. Replace that question mark with a variable. You can pick your favorite letter, but x will do. If $\frac{5}{7} = \frac{x}{32}$ is a proportion, then $7x = 5 \cdot 32$, or $7x = 160$. Then you can find out what x must equal by dividing 160 by 7. Since $160 \div 7 = 22\frac{6}{7}$, you'll need $22\frac{6}{7}$ ounces of yellow paint.

To find the missing piece of a proportion, cross-multiply, and then divide.

EXERCISE

5·3

Find the missing number in each proportion.

1. $\frac{x}{15} = \frac{6}{9}$

2. $\frac{12}{x} = \frac{96}{500}$

3. $\frac{4}{7} = \frac{x}{49}$

4. $\frac{3}{11} = \frac{13}{x}$

5. $\frac{9}{5} = \frac{x}{42}$

6. $\frac{21}{x} = \frac{9}{42}$

7. $\frac{x}{56} = \frac{3}{20}$

8. $\frac{18}{20} = \frac{81}{x}$

9. $\frac{x}{125} = \frac{40}{600}$

10. $\frac{91}{x} = \frac{21}{51}$

Percent

When you work with ratios or fractions, it's easy to compare fractions with the same denominator or fractions whose common denominator is obvious. You know that $\frac{4}{5}$ is bigger than $\frac{3}{5}$, and that

$\dfrac{1}{2}$ is less than $\dfrac{3}{4}$. But which is bigger: $\dfrac{62}{63}$ or $\dfrac{71}{72}$? That comparison is more difficult, and you're probably not crazy about finding a common denominator for 63 and 72. (It's 504, in case you were wondering.) It would be nice if there were a way to compare a lot of different fractions using the same denominator. That's why you turn to percents.

The word *percent* actually means "out of 100." Changing a fraction (or a decimal) into a percent is like changing it to a fraction with a denominator of 100. If all your fractions have a denominator of 100, it's easy to compare them. $\dfrac{65}{100}$ is bigger than $\dfrac{61}{100}$, and $\dfrac{49}{100}$ is smaller than either of them. Another name for $\dfrac{65}{100}$ is 65%, and $\dfrac{61}{100}$ is 61%, literally, 61 out of 100.

Decimal to percent

A decimal with two digits after the decimal point, like 0.75, is 75 hundredths, or $\dfrac{75}{100}$. It's easy to change decimals like this into percents. Seventy-five hundredths is seventy-five out of one hundred, or seventy-five percent, so 0.75 = 75%. In general, to change a decimal to a percent, move the decimal point two places to the right, adding zeros if necessary, and add a percent sign. So 0.683 is 68.3%, and 0.7 is 70%. If the decimal repeats, you can write out the first few digits of the pattern, move the decimal point, and then express the repeating pattern by putting a bar over the repeating digits. The decimal $0.\overline{3} = 0.\underline{3}3333... = 33.\overline{3}\%$.

EXERCISE

5·4

Change each decimal to a percent.

1. 0.45

2. 0.867

3. 0.02

4. 0.0003

5. 0.15$\overline{6}$

6. 0.$\overline{49}$

7. 0.2$\overline{3}$

8. 4.583

9. 2.9

10. 0.1$\overline{23}$

Percent to decimal

Moving in the other direction, changing a percent to a decimal, just requires reversing the steps. To change 16% to a decimal, drop the percent sign, and move the decimal point two places left. 16% = 0.16. When the percent ends with a fraction, that's a sign that the percent came from a repeating decimal. It may be easier to think about moving the decimal point if you rewrite the fraction part as a decimal.

$$16\dfrac{2}{3}\% = \underline{16}.\overline{6}\% = 0.16\overline{6} = 0.1\overline{6}$$

To change a percent to a decimal, drop the percent sign and move the decimal point two places to the left.

Change each percent to a decimal.

1. 29%

2. 43.5%

3. 7%

4. $33\frac{1}{3}$%

5. 54.54%

6. 0.6%

7. 129%

8. 4.59%

9. 45.9%

10. 459%

Fraction to percent

To change a fraction to a percent, use whichever of these two methods is more convenient.

If it's easy to change the fraction to an equal fraction with a denominator of 100, do so. Take the numerator of the fraction with a denominator of 100 and add a percent sign. $\frac{1}{4} = \frac{25}{100} = 25\%$.

Otherwise, change the fraction to a decimal by dividing the numerator by the denominator. Then change the decimal to a percent. Move the decimal point two places to the right (that is, multiply by 100) and add a percent sign. To change $\frac{17}{40}$ to a percent, first change to a decimal.

$$
\begin{array}{r}
0.425 \\
40\overline{)17.000} \\
\underline{160} \\
100 \\
\underline{80} \\
200 \\
\underline{200} \\
\end{array}
$$

Then move the decimal point two places to the right so that 0.425 becomes 42.5, and add a percent sign.

$$\frac{17}{40} = 42.5\%$$

You can also choose to put the remainder from the division over the divisor as a fraction. When you change $\frac{1}{6}$ to a percent, you first change to a decimal.

$$
\begin{array}{r}
0.16 \\
6\overline{)1.00} \\
\underline{6} \\
40 \\
\underline{36} \\
4 \\
\end{array}
$$

You could keep dividing and see that the 6 will repeat, or you could express $\frac{1}{6}$ as $0.16\frac{4}{6} = 0.16\frac{2}{3}$ and then change to $16\frac{2}{3}\%$.

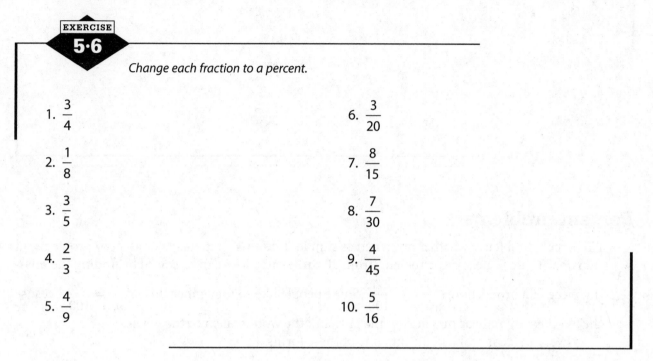

Change each fraction to a percent.

1. $\frac{3}{4}$

2. $\frac{1}{8}$

3. $\frac{3}{5}$

4. $\frac{2}{3}$

5. $\frac{4}{9}$

6. $\frac{3}{20}$

7. $\frac{8}{15}$

8. $\frac{7}{30}$

9. $\frac{4}{45}$

10. $\frac{5}{16}$

Percent to fraction

To change a percent to a fraction, you can usually rely on its name. *Percent* means "out of one hundred," so to change 38% to a fraction, you can just put 38 over 100 and simplify.

$$38\% = \frac{38}{100} = \frac{19}{50}$$

If the percent involves a fraction, change it to a decimal. The percent $45\frac{1}{9}\%$ becomes $45.\overline{1}\%$. Drop the percent sign, and move the decimal point two places left, to change the percent to a decimal. $45.\overline{1}\%$ becomes $0.45111....$ Change the decimal to a fraction and simplify. $0.45111...$ becomes $\frac{451 - 45}{900} = \frac{406}{900} = \frac{203}{450}$.

Here are more examples:

$$17\frac{1}{4}\% = 17.25\% = 0.1725 = \frac{1725}{10,000} = \frac{69}{400}$$

$$8.5\% = 0.085 = \frac{85}{1,000} = \frac{17}{200}$$

Change each percent to a fraction.

1. 15%

2. 120%

3. 2%

4. 0.75%

5. $66.\overline{6}\%$

6. $4\frac{4}{9}\%$

7. 0.1%

8. $0.\overline{1}\%$

9. 1%

10. 100%

Percent problems

All percent problems, whether presented as simple "find this" statements or as story problems, fit into one of three categories. Problems from all three categories can be solved by finding the missing piece of a proportion: $\dfrac{part}{whole} = \dfrac{\%}{100}$. Some people like to remember this as $\dfrac{is}{of} = \dfrac{\%}{100}$ because the word *is* generally occurs next to the *part* and the word *of* next to the *whole*.

Every percent problem fits into one of these forms:

◆ Find the percent: $\underbrace{80}_{Part}$ is $\underbrace{\text{what percent}}_{\%}$ of $\underbrace{240}_{Whole}$?

$$\frac{part}{whole} = \frac{\%}{100}$$

$$\frac{80}{240} = \frac{x}{100}$$

$$240x = 8{,}000$$

$$x = \frac{8{,}000}{240} = 33\frac{1}{3}$$

80 is $33\frac{1}{3}\%$ of 240.

◆ Find the part: $\underbrace{\text{What}}_{Part}$ is $\underbrace{60\%}_{\%}$ of $\underbrace{300}_{Whole}$?

$$\frac{part}{whole} = \frac{\%}{100}$$

$$\frac{x}{300} = \frac{60}{100}$$

$$100x = 18{,}000$$

$$x = \frac{18{,}000}{100} = 180$$

180 is 60% of 300.

◆ Find the whole: $\underset{\text{Part}}{\underline{15}}$ is $\underset{\%}{\underline{75\%}}$ of $\underset{\text{Whole}}{\underline{\text{what number}}}$?

$$\frac{part}{whole} = \frac{\%}{100}$$

$$\frac{15}{x} = \frac{75}{100}$$

$$75x = 1{,}500$$

$$x = \frac{1{,}500}{75} = 20$$

15 is 75% of 20.

Percent change

Problems often talk about the percent increase or the percent decrease of a certain quantity. These problems can be solved in the same way as basic percent problems if you just remember that the change is the part, and the original amount is the whole.

Tim and Sarah both bought stock for $25 per share. When Tim sold his stock, the price was $40 per share. When Sarah sold her stock, the price was $15 per share. What was the percent change in each investment?

Tim's investment increased in value from $25 to $40, a change of $40 − $25 = $15. The change of $15 is the part and the original price of $25 is the whole.

$$\frac{part}{whole} = \frac{\%}{100}$$

$$\frac{15}{25} = \frac{x}{100}$$

$$25x = 1{,}500$$

$$x = \frac{1{,}500}{25} = 60$$

Tim's investment increased 60%.

Sarah's investment decreased in value from $25 to $15, a change of $10. The change of $10 is the part and the original price of $25 is the whole.

$$\frac{part}{whole} = \frac{\%}{100}$$

$$\frac{10}{25} = \frac{x}{100}$$

$$25x = 1{,}000$$

$$x = \frac{1{,}000}{25} = 40$$

Sarah's investment decreased 40%.

Solve each problem.

1. 14 is what percent of 70?

2. What is 15% of 160?

3. 27 is 80% of what number?

4. Find 2% of 1,492.

5. 254 is what percent of 500?

6. 95% of 849 is what number?

7. Find 250% of 78.

8. What number represents 0.25% of 200?

9. If a vehicle decreases in value from $30,000 to $23,500, what is the percent decrease?

10. An investment originally worth $50,000 had a 27% increase. Find the change in value.

11. If the bill in a restaurant is $38, and you want to leave a 15% tip, how much should you leave as a tip?

12. If the government takes 28% of your income and you earn $75,000 a year, how much of your income goes to the government?

13. If you buy a stock for $80 per share and sell it for $100 per share, what is the percent increase?

14. A car, purchased for $20,000, loses 14% of its value in the first year. What is the car worth after one year?

15. You're planning a vacation and are considering a hotel room that is advertised at $120 per night. You notice in the fine print that there will be an additional 18% tax on the room. How much tax will you pay? How much will you actually be charged per night, including the tax?

Exponents and roots

To add the same number several times, you can use multiplication as a shortcut. $5 + 5 + 5 + 5 + 5 + 5 = 6(5) = 30$. To multiply the same number several times, as in $2 \times 2 \times 2 \times 2 \times 2 \times 2 \times 2$, you really can't find a shortcut for the work, but you can shorten the way it's written by using an exponent: $2 \times 2 \times 2 \times 2 \times 2 \times 2 \times 2 = 2^7$. The small, raised 7 is the **exponent**, which tells you that you need to multiply 2, the base, seven times. An expression involving an exponent is commonly called a **power**, and the expression 2^7 is commonly read "two to the seventh power."

$$2 \times 2 \times 2 \times 2 \times 2 \times 2 \times 2 = 2^7 = 128$$

Evaluating

The exponents you'll see most often are the second power, or **square**, and the third power, or **cube**. Those powers take their common names from the geometric figures square and cube. If you have a square in which each side measures 8 inches, the area of the square is 8^2. The volume of a cube that measures 7 by 7 by 7 is 7^3.

Exponents are a short way of writing repeated multiplication, but there is no shortcut for doing the multiplication. To find the value of 3^5, you have to multiply $3 \times 3 \times 3 \times 3 \times 3$. Trying to invent a shortcut, like multiplying the base and the exponent, is a common mistake, but it only gets you into trouble. You need to multiply:

$$
\begin{aligned}
3^5 &= 3 \times 3 \times 3 \times 3 \times 3 \\
&= 9 \times 3 \times 3 \times 3 \\
&= 27 \times 3 \times 3 \\
&= 81 \times 3 \\
&= 243
\end{aligned}
$$

When you raise a negative number to a power, watch your signs. A negative number raised to an even power will be positive, but a negative number raised to an odd power will be negative.

$$
\begin{aligned}
(-2)^4 &= (-2) \times (-2) \times (-2) \times (-2) \\
&= 4 \times (-2) \times (-2) \\
&= -8 \times (-2) \\
&= 16
\end{aligned}
\qquad
\begin{aligned}
(-3)^3 &= (-3) \times (-3) \times (-3) \\
&= 9 \times (-3) \\
&= -27
\end{aligned}
$$

To raise a negative number to a power, always put the negative number in parentheses.

$$(-2)^4 = (-2) \times (-2) \times (-2) \times (-2) \text{ but } -2^4 = -(2^4) = -(2 \times 2 \times 2 \times 2) = -16$$

EXERCISE
6·1

Evaluate each power.

1. 6^2

2. 4^3

3. $(-1)^{10}$

4. $(-2)^2$

5. $\left(\dfrac{3}{4}\right)^2$

6. -6^2

7. $(-5)^3$

8. 2^{10}

9. 2.5^4

10. $\left(\dfrac{1}{2}\right)^3$

Multiplying

When you multiply two or more powers, you certainly have the option to evaluate each power and then multiply. That may be the best way to tackle many problems.

$$4^2 \times 2^5 = 16 \times 32 = 512$$

When you're multiplying powers of the same base, however, there's a shortcut available to you. Think about a problem like $3^4 \times 3^3$. You could do it as 81×27, but if you write out what the powers mean, you have $(3 \times 3 \times 3 \times 3) \times (3 \times 3 \times 3)$, which is clearly a power of 3. It's 3^7, because you have 4 threes times 3 threes—a total of 7 threes.

To multiply powers of the same base, keep the base and add the exponents.

$$3^4 \times 3^3 = 3^{4+3} = 3^7$$

EXERCISE
6·2

Perform each multiplication by keeping the base and adding the exponents, if possible.
Leave your answer as a power.

1. $2^3 \times 2^5$

2. $3^4 \times 3^7$

3. $9^5 \times 9^8$

4. $12^4 \times 12^2$

5. $x^2 \cdot x^3$

6. $a^5 \times a^7$

7. $2^5 \times 3^5 \times 3^8$

8. $a^6 \times b^2 \times a^9$

9. $x^7 \cdot x^5$

10. $z^{32} \cdot z^{15}$

Dividing

When you divide powers, your choices are similar to those you face when multiplying. When dividing, you can evaluate each power and then divide, but if the bases are the same, you can take advantage of shortcuts. Because you're dividing powers of the same base, you can do lots of canceling.

$$\frac{5^7}{5^3} = \frac{\cancel{5} \times \cancel{5} \times \cancel{5} \times 5 \times 5 \times 5 \times 5}{\cancel{5} \times \cancel{5} \times \cancel{5}} = 5^4$$

The second option is simpler. To divide powers of the same base, keep the base and subtract the exponents.

$$7^{12} \div 7^9 = 7^{12-9} = 7^3$$

EXERCISE 6·3

Perform each division by keeping the base and subtracting the exponents, if possible. Leave your answer as a power.

1. $2^5 \div 2^3$

2. $3^7 \div 3^4$

3. $9^8 \div 9^5$

4. $12^5 \div 12^3$

5. $\dfrac{x^5}{x^2}$

6. $\dfrac{a^7}{a^5}$

7. $\dfrac{3^5 \times 3^8}{3^{11}}$

8. $\dfrac{a^6 \times a^7}{a^9}$

9. $x^7 \div x^5$

10. $z^{32} \div z^{15}$

Powers of powers

When you raise a number to a power, you do a multiplication, in which the base is the only factor and the exponent tells you how many times to use it as a factor. 7^4 tells you to multiply with 7 as a factor four times: $7 \times 7 \times 7 \times 7$. When you raise a power, like 2^3, to a power, you're saying you want to use the power 2^3 as a factor a certain number of times.

$$(2^3)^5 = 2^3 \times 2^3 \times 2^3 \times 2^3 \times 2^3$$

You know that when you multiply powers of the same base, you add the exponents. To raise a power to a power, keep the base and multiply the exponents.

$$(2^3)^5 = 2^3 \times 2^3 \times 2^3 \times 2^3 \times 2^3 = 2^{3+3+3+3+3} = 2^{3 \times 5} = 2^{15}$$

$$(t^5)^7 = t^{5 \times 7} = t^{35}$$

Use the rules for powers to put each expression in simplest form.

1. $(4^3)^2$ 6. $(x^3)^4$

2. $(5^2)^3$ 7. $(a^2)^9$

3. $(7^3)^4$ 8. $(y^8)^2$

4. $((-2)^3)^5$ 9. $(b^3)^8$

5. $(21^2)^4$ 10. $(v^4)^4$

Powers of products and quotients

When you raise a product, such as 4×5 or $2a$, to a power, you use the entire product as a factor several times. $(4 \times 5)^2 = (4 \times 5) \times (4 \times 5) = 4 \times 4 \times 5 \times 5 = 4^2 \times 5^2$. Each number in the product is affected by the exponent.

To raise a product to a power, raise each factor to that power.

$$(2a)^3 = 2^3 a^3$$

A quotient, such as $12 \div 7$, can be written as a fraction, $\dfrac{12}{7}$. When you raise a fraction to a power, you multiply the numerators and multiply the denominators.

$$\left(\frac{12}{7}\right)^2 = \frac{12}{7} \times \frac{12}{7} = \frac{12^2}{7^2}$$

To raise a quotient, or fraction, to a power, raise both the numerator and the denominator to that power.

Use the rules for power of a product and power of a quotient to simplify each expression.

1. $(4 \times 2)^2$ 6. $(2 \div 4)^3$

2. $(-3 \times 10)^2$ 7. $\left(\dfrac{7}{10}\right)^3$

3. $(2x^2)^4$ 8. $\left(\dfrac{4x^5}{8x^3}\right)^2$

4. $(3x^2)^3$ 9. $\left(\dfrac{3x^7}{5x^4}\right)^2$

5. $(-2a^3bc^4)^3$ 10. $\left(\dfrac{6a^2bc^4}{3abc}\right)^5$

Special exponents

Once you're acquainted with exponents and how they work, it's a good idea to take a moment to look at a few unusual exponents.

An exponent of 1, on any base, produces the base. You use the base as a factor once, so you just get the base. So $4^1 = 4$ and $107^1 = 107$. You don't generally show an exponent of 1, but you sometimes get one, for example, $6^9 \div 6^8 = 6^1 = 6$.

Another unusual exponent you run into when you divide is the zero exponent. According to the rule that says we should keep the base and subtract the exponents when we divide, $\dfrac{5^4}{5^4} = 5^0$. From arithmetic, you know that any number divided by itself equals 1, so $\dfrac{5^4}{5^4} = 1$. That means 5^0 must equal 1. Any non-zero number, raised to the zero power, equals one. (Zero to the zero power is indeterminate, because two conflicting rules would apply. On one hand, zero to any power should be zero, and on the other, any number to the zero power should be one.)

If you follow the division rule, $8^3 \div 8^4$ should equal $8^{3-4} = 8^{-1}$. If you look at that division written out, you can see that the result is $\dfrac{\cancel{8} \times \cancel{8} \times \cancel{8}}{\cancel{8} \times \cancel{8} \times \cancel{8} \times 8} = \dfrac{1}{8}$. A number to the -1 power is equal to the reciprocal of the number. $8^{-1} = \dfrac{1}{8}$, $382^{-1} = \dfrac{1}{382}$, and $\left(\dfrac{4}{5}\right)^{-1} = \dfrac{5}{4}$.

When you combine the definition of a -1 exponent with the power of a power rule, you can define any negative exponent. $(5^{-1})^3 = \left(\dfrac{1}{5}\right)^3 = \dfrac{1}{5} \times \dfrac{1}{5} \times \dfrac{1}{5} = \dfrac{1}{5^3} = 5^{-3}$. An exponent of -1 produces the reciprocal of the base. When you see any other negative exponent, find the positive power and then find its reciprocal. $8^2 = 64$ and $8^{-2} = \dfrac{1}{64}$, the reciprocal of 64. A number to the $-n$ power is equal to the reciprocal of the nth power of the base.

EXERCISE 6·6

Simplify each expression. Assume that all variables represent positive numbers.

1. $(2 \times 4)^{-1}$

2. $\left(\dfrac{2}{3}\right)^{-3}$

3. $y^{-2}\, y^3\, y^{-4}$

4. $(x^4\, y^{-4})^0\, (2x^{-2})$

5. $\dfrac{12x^5 y^8}{4x^{-2} y^{-4}}$

6. $(-2x^{-1}\, y^{-3})^3$

7. $(4x^3\, y^{-45})^4\, (4x^3\, y^{-45})^{-4}$

8. $\left(\dfrac{2x^3}{z^4}\right)^{-3}$

9. $\dfrac{15x^{-5} y^2}{2x^5 y} \cdot \dfrac{xy^{-2}}{5x^3 y^{-2}}$

10. $\left(\dfrac{3x^2 y}{2x^{-3} y^2}\right)^{-2}$

Large and small

In the ancient Egyptian system of numeration, the symbol for one million was a hieroglyph described as a man kneeling with his arms spread in amazement. For that society, a million was an awe-inspiring number, so large it was hard to imagine. In contemporary society, where telescopes and microscopes allow us to deal commonly with immense distances and infinitesimally tiny objects, we need more flexible and powerful ways to deal with size.

Rounding and estimation

You might look up the fact that the average distance from the earth to the moon is 238,857 miles and that the speed of the space shuttle, when in orbit, was 17,285 miles per hour. You could get your calculator out and determine that, if you could travel at that speed, it would take an average of $13\frac{14,152}{17,285}$ hours, or approximately 13.8187445762 hours, to reach the moon. On the other hand, if you could travel at the speed the shuttle was lifted into orbit by the solid booster rockets—3,094 miles per hour—the trip would take $77\frac{619}{3,094}$ hours, or approximately 77.20006464 hours. (The second is the more reasonable, by the way.)

If those awkward fractions and long decimals left you feeling that 13 or 77 was as much information as you really needed, then like most people, you recognize that often an estimate is adequate, especially when you're dealing with very large or very small numbers. To arrive at a reasonable estimate, you'll want to round the numbers.

You might round the average distance from the earth to the moon, which is 238,857 miles, to 200,000 miles. That says that you want to focus on the largest decimal place, the hundred-thousands, and that 238,857 is closer to 200,000 than to 300,000. You might round it to the nearest ten thousand, by deciding if it's closer to 230,000 or 240,000, the "round" numbers below it or above it. To round a number to a particular place, look to the digit following that place. If the digit is 5 or more, round up. If it's 4 or less, round down. The number 238,857 rounds up to 240,000 because the 8 in the thousands place is greater than 5, telling you that 238,857 is closer to 240,000 than 230,000. The number 3,094, if rounded to the nearest thousand would be 3,000 because the digit following the thousands place is a 0, telling you to round down.

The average distance from earth to the moon rounded to the nearest ten thousand is 240,000 miles and the orbital speed of the shuttle, 17, 285 miles per hour, rounded to the nearest ten thousand is 20,000. With those estimates, you can easily divide 240,000 ÷ 20,000 and get an estimated travel time of 12 hours. That estimate is a bit less than the actual calculation, but close enough for most needs.

Scientific notation

Scientific notation is a method of expressing very large or very small numbers in a compact form. It makes use of exponents and our decimal place value system. Each place in our system represents a value that is a power of 10, so every number can be written as a small number times a power of 10. The number 4,000 can be written as $4 \times 1,000$ or 4×10^3. The number 0.000005 can be written as 5×0.000001 or 5×10^{-6}.

While there are many different ways that you could express the same number, the agreement is that scientific notation writes a number as the product of a number greater than or equal to 1 but less than 10, and a power of 10. For numbers greater than 1, the power of 10 has a positive exponent.

$$4,670,000 = 4.67 \times 10^6$$

$$93,000,000,000 = 9.3 \times 10^{10}$$

$$84,620 = 8.462 \times 10^4$$

To change a large number in standard form to scientific notation:

1. Count the number of digits in the number, and subtract 1. This will be the exponent.

2. Place the decimal point after the first digit and drop trailing zeros.

3. Multiply by 10 with the exponent you found in step 1.

Count the digits in 4,300,000 and you get 7. Subtract 1 and your exponent will be 6. Put the decimal point after the 4 and drop the zeros, so you have 4.3. Multiply by 10^6. So $4,300,000 = 4.3 \times 10^6$. The exponent of 6 tells you to move the decimal point six places to the right.

$$4.3 \times 10^6 = 4,\underbrace{300,000}_{\text{six places right}}$$

When you change a number less than 1 to scientific notation, you're dealing with a fraction, so the exponent on the 10 will be a negative number.

$$0.038 = 3.8 \times 10^{-2}$$

$$0.00072 = 7.2 \times 10^{-4}$$

$$0.000000000005 = 5 \times 10^{-12}$$

To change a small number in standard form to scientific notation:

1. Count the number of zeros after the decimal point but before the first non-zero digit, and add 1. Make the number negative. This will be the exponent.

2. Place the decimal point after the first non-zero digit and drop leading and trailing zeros.

3. Multiply by 10 with the exponent you found in step 1.

To change 0.0000784 to scientific notation, count the four zeros after the decimal point but before the 7. Add 1 to get 5, then make it negative (-5). Place the decimal point after the 7 and drop the zeros before the 7 to get 7.84. Multiply by 10^{-5}. $0.0000784 = 7.84 \times 10^{-5}$. The exponent of -5 tells you to move the decimal point five places to the left.

$$7.84 \times 10^{-5} = \underbrace{.00007}_{\text{five places left}} 84$$

The advantage of expressing very large or very small numbers in scientific notation is that it allows you to use the properties of exponents to calculate easily.

To multiply or divide numbers in scientific notation, multiply or divide the decimals using standard arithmetic. Multiply the powers of 10 by adding the exponents, or divide them by subtracting exponents. Check the result of the multiplication to see if it is actually in scientific notation. The multiplication or division of decimals can produce a result that is not a number between 1 and 10. If that happens, you'll need to adjust so that the answer is in scientific notation.

To multiply $(2.5 \times 10^{13}) \times (6.4 \times 10^{11})$, multiply 2.5×6.4 to get 16.00 and multiply $10^{13} \times 10^{11} = 10^{24}$. Your preliminary result is 16×10^{24}, but that's not scientific notation because 16 is greater than 10.

$$16 \times 10^{24} = 1.6 \times 10 \times 10^{24} = 1.6 \times 10^{25}$$

To divide $(2.4 \times 10^{-8}) \div (9.6 \times 10^{-12})$, divide $2.4 \div 9.6$ to get 0.25, and notice that 0.25 is less than 1. Divide $10^{-8} \div 10^{-12} = 10^{-8-(-12)} = 10^4$. Adjust so that the quotient is in scientific notation.

$$(2.4 \times 10^{-8}) \div (9.6 \times 10^{-12}) = 0.25 \times 10^4 = 2.5 \times 10^{-1} \times 10^4 = 2.5 \times 10^3$$

For questions 1 through 5, round each number to the indicated place, and use the rounded versions to estimate the result of the calculation.

1. ten thousands: 83,681 + 119,392

2. millions: 9,384,583 − 5,639,299

3. thousands: 62,973 ÷ 8,856

4. hundred thousands: 293,449 × 781,772

5. millions: 11,735,882 × 49,948,221

For questions 6 through 8, write each number in standard form.

6. 9.29×10^6

7. 1.35×10^{-3}

8. 4.73×10^{12}

For questions 9 through 11, write each number in scientific notation.

9. 0.000234

10. 87,000

11. 972,734,000,000

For questions 12 through 15, evaluate each expression and write your answer in scientific notation.

12. $(5 \times 10^{-3})(3 \times 10^6)$

13. $\dfrac{7.7 \times 10^{-4}}{1.1 \times 10^{-1}}$

14. $(5 \times 10^{-2})^3$

15. $(4.2 \times 10^{-8})(3 \times 10^5)$

Roots

Every operation has an opposite, or inverse. The opposite of addition is subtraction and the inverse of multiplication is division. The opposite of exponentiation (the operation of raising a number to a power) is the process of taking a root. If $6^2 = 36$, then the square root of 36 is 6. The symbol for square root is called a **radical**, and the number under the radical sign is called the **radicand**. In the statement $\sqrt{36} = 6$, 36 is the radicand. Every power has a corresponding root, but the basic symbol remains the radical. For roots other than square roots, you just add a little number, called an index, in the crook of the sign. If $4^3 = 64$, then $\sqrt[3]{64} = 4$. This is called the cube root. $\sqrt[4]{16}$ is the fourth root of 16. $\sqrt[10]{1,024}$ is the tenth root of 1,024. Don't get frightened by that, however. Most of the roots you encounter are square roots.

Evaluating and simplifying

Finding the value of a root is a job that often requires a calculator (or, in the days before calculators, a table of roots). Just think about the squares you know. $5^2 = 25$ and $6^2 = 36$, so $\sqrt{25} = 5$ and $\sqrt{36} = 6$, but that means that all the numbers between 25 and 36 have square roots that are not whole numbers. In fact, many square roots are irrational numbers. Rather than dealing with all those decimals, you'll want to put the square roots in simplest radical form. Because radicals are numbers, they can be added, subtracted, multiplied, and divided, and that means you'll sometimes see fractions involving radicals. It also means that you'll want the radicals in the simplest possible form to make your work easier.

A square root is in its simplest form when:

- The number under the radical has no perfect square factors
- There is no radical in the denominator of a fraction

To put a radical in simplest form, express the **radicand**, the number under the radical, as a product of two numbers, at least one of which is a perfect square. Take the time to look for the largest perfect square factor possible. For example, simplify the square root of 48. $48 = 4 \times 12$ and 4 is a perfect square, but 3×16 also equals 48, and 16 is a larger perfect square. You can look for these factors by using a factor tree or by dividing the radicand by small divisors and checking to see if the quotient is a perfect square. Once you have the factors, give each factor its own radical. $\sqrt{48} = \sqrt{3 \times 16} = \sqrt{3} \times \sqrt{16}$. Then take the square root you know. $\sqrt{3} \times \sqrt{16} = \sqrt{3} \times 4 = 4\sqrt{3}$. The simplest radical form of $\sqrt{48}$ is $4\sqrt{3}$. This is usually read as 4 radical 3 or 4 root 3.

Sometimes when you simplify a radical, you discover that the radicand was actually a perfect square. $\sqrt{2,304}$ isn't a square root you'd know from memory, but when you start to simplify it, you discover that 2,304 is a perfect square.

$$\sqrt{2,304} = \sqrt{4 \times 576} = \sqrt{4 \times 4 \times 144} = \sqrt{4} \times \sqrt{4} \times \sqrt{144} = 2 \times 2 \times 12 = 48$$

When you take the square root of a fraction, you take the square root of the numerator and the square root of the denominator. (Does that remind you of the rule for exponents? It should. There's actually a way to express roots as exponents, but that's for another time.)

When finding the simplest form, you don't want to leave any radicals in the denominator. Sometimes that's easy. For example, you know $\sqrt{\dfrac{3}{4}} = \dfrac{\sqrt{3}}{\sqrt{4}}$ and you know that $\sqrt{4} = 2$, so $\sqrt{\dfrac{3}{4}} = \dfrac{\sqrt{3}}{\sqrt{4}} = \dfrac{\sqrt{3}}{2}$. There is no radical in the denominator, so you've found the simplest form. But look at $\sqrt{\dfrac{1}{3}}$.

You can figure out that $\sqrt{\dfrac{1}{3}} = \dfrac{\sqrt{1}}{\sqrt{3}} = \dfrac{1}{\sqrt{3}}$, but you don't want to leave that radical in the denominator. To eliminate it, you **rationalize the denominator**, which means you multiply the numerator and the denominator by $\sqrt{3}$. You'll be multiplying by $\dfrac{\sqrt{3}}{\sqrt{3}} = 1$, so the value won't change, but the look of the fraction will, and when you multiply a square root by itself, that is, when you square the square root, the squaring and the square root cancel one another. We say the radical lifts, and you get the number under the radical.

$$\frac{1}{\sqrt{3}} \cdot \frac{\sqrt{3}}{\sqrt{3}} = \frac{\sqrt{3}}{3}$$

EXERCISE 6·8

Put each expression in simplest radical form.

1. $\sqrt{128}$

2. $\sqrt{675}$

3. $\sqrt{180}$

4. $\sqrt{1,225}$

5. $\dfrac{1}{\sqrt{2}}$

6. $\dfrac{12}{\sqrt{6}}$

7. $\dfrac{2\sqrt{5}}{5\sqrt{2}}$

8. $\sqrt{75x^2y^4}$

9. $\sqrt{80a^3b^4c^2}$

10. $\sqrt{\dfrac{120x^2y}{15xy^3}}$

Multiplying and dividing

To multiply or divide radical expressions, the rule you want to remember is inside with inside, and outside with outside. To multiply $6\sqrt{2} \times 5\sqrt{3}$, multiply the numbers that are inside radicals, and keep the result inside a radical, and multiply the numbers that are outside the radicals and put the result outside.

$$6\sqrt{2} \times 5\sqrt{3} = 6 \times 5 \times \sqrt{2} \times \sqrt{3} = 30\sqrt{2 \times 3} = 30\sqrt{6}$$

Sometimes this will give you a result that has a radical that needs simplifying. $2\sqrt{6} \times 5\sqrt{3} = 10\sqrt{18}$, but that's not in simplest form. You need to simplify the radical.

$$2\sqrt{6} \times 5\sqrt{3} = 10\sqrt{18} = 10\sqrt{9 \times 2} = 10 \times \sqrt{9} \times \sqrt{2} = 10 \times 3 \times \sqrt{2} = 30\sqrt{2}$$

Follow the same inside/outside rule when you divide radicals. $\dfrac{14\sqrt{6}}{7\sqrt{2}} = \dfrac{14}{7} \times \sqrt{\dfrac{6}{2}} = 2\sqrt{3}$. If trying to divide is going to give you decimals under the radical, just rationalize the denominator.

$$\frac{27\sqrt{5}}{3\sqrt{7}} = \frac{27}{3} \times \sqrt{\frac{5}{7}} = 9\sqrt{\frac{5}{7}} = 9\sqrt{\frac{5}{7}} \times \frac{\sqrt{7}}{\sqrt{7}} = 9 \times \frac{\sqrt{35}}{7} = \frac{9}{7}\sqrt{35}$$

Perform each multiplication or division and give your answers in simplest radical form.

1. $\sqrt{12} \cdot \sqrt{27}$

2. $\sqrt{5} \cdot \sqrt{35}$

3. $2\sqrt{3} \cdot \sqrt{6}$

4. $4\sqrt{30} \cdot 5\sqrt{33}$

5. $-5\sqrt{45} \cdot 8\sqrt{75}$

6. $\dfrac{18\sqrt{15}}{6\sqrt{5}}$

7. $\dfrac{9\sqrt{162}}{27\sqrt{8}}$

8. $\dfrac{32\sqrt{75}}{8\sqrt{15}}$

9. $\dfrac{10\sqrt{6}}{\sqrt{8}}$

10. $\dfrac{7\sqrt{288}}{21\sqrt{48}}$

Adding and subtracting

Adding and subtracting radicals is a bit like adding and subtracting fractions. You need to have a common denominator to add or subtract fractions, and you must have like radicals to add or subtract. When you write $5\sqrt{3}$, you're saying you have 5 of these numbers called "radical three." If you add $5\sqrt{3} + 7\sqrt{3}$, you have 5 of these things plus 7 of the same thing, so you have $12\sqrt{3}$, or 12 of those things. But if you try to add $5\sqrt{3} + 7\sqrt{2}$, you don't have 12 of anything. You have 5 of one thing and 7 of another.

To add or subtract like radicals, add or subtract the numbers in front, and keep the radical.

$$15\sqrt{7} - 9\sqrt{7} = 6\sqrt{7}$$

Before you decide that you don't have like radicals, and so can't add or subtract, and move on to other things, take a moment to make sure both radicals are in simplest form. Radicals that may not look alike before simplifying may turn out to be like radicals when you see them in simplest form.

$$
\begin{aligned}
3\sqrt{50} + 4\sqrt{98} &= 3 \times \sqrt{25 \times 2} + 4 \times \sqrt{49 \times 2} \\
&= 3 \times 5 \times \sqrt{2} + 4 \times 7 \times \sqrt{2} \\
&= 15\sqrt{2} + 28\sqrt{2} \\
&= 43\sqrt{2}
\end{aligned}
$$

Simplify radicals and perform each addition or subtraction, if possible.

1. $\sqrt{8}+\sqrt{72}$

2. $\sqrt{12}+2\sqrt{48}$

3. $2\sqrt{75}+9\sqrt{147}$

4. $\sqrt{18}-3\sqrt{50}+5\sqrt{8}$

5. $12\sqrt{405}+7\sqrt{500}-3\sqrt{180}$

6. $\dfrac{12}{\sqrt{6}}+\sqrt{6}$

7. $\dfrac{4}{2\sqrt{2}}+\dfrac{7\sqrt{2}}{4}$

8. $\dfrac{8\sqrt{2}}{6\sqrt{3}}+\dfrac{7\sqrt{3}}{9\sqrt{2}}$

9. $\dfrac{4}{3\sqrt{5}}-\dfrac{2}{\sqrt{45}}$

10. $\dfrac{6\sqrt{7}}{2\sqrt{5}}-\dfrac{7\sqrt{5}}{4\sqrt{7}}$

Equations and inequalities ◆·7·

A **variable** is a letter (or other symbol) that stands for a value that can change or a number that is unknown. If you want to say that there is some number that can be added to 12 to make 33, you could write $12 + x = 33$ or $12 + \square = 33$. The x or the empty box is the variable because it takes the place of the number. The introduction of variables is where arithmetic begins to shift to algebra.

Once algebra introduces variables, the quest to know the unknown begins. Solving equations and inequalities is the process of determining what value (or values) could replace the variable and make the statement true. The methods of solving equations and inequalities are similar. Let's look at equations first.

Solving linear equations

Linear equations take their name from the fact that their graphs are lines. They are equations that contain only the first power of one variable. No exponents appear on variables and no variables appear under radicals. Solving the equation means determining the value that replaces the variable to make a true statement, and that's accomplished by doing the opposite of what you see.

One-step equations

If you know someone added 4 to a number and got 7, you can work backward to find out that the number was $7 - 4$, or 3. That situation is represented by the equation $y + 4 = 7$, and you can find the value of y that makes the equation true by subtracting 4 to undo the addition. Officially, you subtract 4 from both sides of the equation.

$$y + 4 - 4 = 7 - 4$$
$$y = 3$$

If an equation was formed by adding, you solve it by subtracting. If it was formed by subtracting, you solve it by adding. The equation $p - 7 = 2$ can be solved by adding 7 to both sides.

$$p - 7 + 7 = 2 + 7$$
$$p = 9$$

The key to solving any equation is getting the variable all alone on one side of the equation. You isolate the variable by performing inverse, or opposite, operations.

If the variable has been multiplied by a number, solve the equation by dividing both sides by that number.

$$-5x = 35$$

$$\frac{\cancel{-5}x}{\cancel{-5}} = \frac{35}{-5}$$

$$x = -7$$

If the variable has been divided by a number, multiply both sides by that number to find the value of the variable.

$$\frac{x}{6} = -2$$

$$\frac{x}{\cancel{6}} \cdot \cancel{6} = -2 \cdot 6$$

$$x = -12$$

EXERCISE

7·1

Solve each equation by adding or subtracting.

1. $x + 7 = 12$

2. $y - 5 = 16$

3. $t + 5 = 16$

4. $w - 13 = 34$

5. $x + \dfrac{1}{2} = \dfrac{7}{2}$

6. $z - 2.9 = 3.1$

7. $y - 4\dfrac{3}{4} = 7\dfrac{1}{2}$

8. $x + 10 = 6$

9. $y - 12 = -4$

10. $t + 11 = -4$

Solve each equation by multiplying or dividing.

11. $8x = 48$

12. $\dfrac{z}{7} = 8$

13. $-6y = 42$

14. $\dfrac{t}{9} = -7$

15. $15x = 45$

16. $\dfrac{w}{11} = 23$

17. $\dfrac{4}{5}t = \dfrac{8}{15}$

18. $\dfrac{m}{4.3} = -3.1$

19. $-1.4x = 4.2$

20. $\dfrac{z}{7} = -35$

Two-step equations

While the single inverse operation is the key to solving equations, most equations require two or more operations to find a solution. The equation $4x - 5 = 19$ says that if you start with a number, x, multiply it by 4 and then subtract 5, the result is 19. To solve for x, you will need to perform the opposite, or inverse, operations in the opposite order. You are undoing, stripping away what was done to x, and working your way back to where things started. Undo the subtraction by adding 5.

$$4x - 5 = 19$$
$$4x = 19 + 5$$
$$4x = 24$$

Then undo the multiplication by dividing by 4.

$$4x = 24$$
$$x = \frac{24}{4}$$
$$x = 6$$

EXERCISE

7·2

Solve each equation.

1. $3x - 11 = 37$

2. $-8t + 7 = 23$

3. $4 - 3x = -11$ (Hint: rewrite as $-3x + 4 = -11$ if that's easier)

4. $5 + 3x = 26$

5. $\dfrac{x}{4} - 8 = -5$

6. $-5x + 19 = -16$

7. $\dfrac{x}{3} + 7 = 16$

8. $13 - 4x = -15$

9. $\dfrac{x}{2} - \dfrac{3}{4} = \dfrac{1}{4}$

10. $12x - 5 = -53$

Terms and expressions

In arithmetic, you learn to work with numbers, and then as you move toward algebra, you introduce variables, but when you start to combine numbers and variables, what do you call the result? The word **term** is used to describe numbers or variables or products of numbers and variables. -4 qualifies as a term, because it's a number, and x is a variable, so it's a term, and so is $-4x$ or $3.5y^2$ or $\dfrac{3}{5}xy$. The powers are allowed because they're shorthand for repeated multiplication, and the fraction is OK because it's a number. Any number, variable, or product of numbers and variables is a term.

Once you start to add or subtract, however, you begin to form what are generally called **expressions**. Examples of expressions include $4x^2 + 3xy$ and $t^4 - 3t(5t + 2)$ and many others. Some expressions have properties that allow you to apply other labels, like **polynomial**, but in general, when you add or subtract terms, you form expressions. You always want to make expressions as simple as possible, and the principal ways you do that are removing parentheses and combining like terms.

Removing parentheses

Parentheses (and other grouping symbols) appear in expressions for one of three reasons. Sometimes, the parentheses don't serve any real mathematical purpose. They're just used to help you understand what the expression is talking about. If you wanted to add the expression $4x^2 + 3xy$ to the expression $5x - 3x^2 y$, you could just write $4x^2 + 3xy + 5x - 3x^2 y$ but you might choose to write $(4x^2 + 3xy) + (5x - 3x^2 y)$ just to make it clearer that you were adding those two pieces. Parentheses that don't have any mathematical purpose can just be dropped.

The principal use of parentheses that does have a mathematical purpose is to tell you that you need to multiply a term times an expression, as in $3t(5t + 2)$, or an expression times an expression, as in $(x + 5)(x - 3)$. To eliminate those parentheses, you need to do the multiplication, using the **distributive property**.

The distributive property is a principle of arithmetic that tells you that when you have to multiply a number times a sum, like $4(8 + 3)$, you'll get the same answer whether you add first, then multiply— $4(8 + 3) = 4 \times 11 = 44$—or multiply each of the addends by 4, and then add: $4(8 + 3) = 4 \times 8 + 4 \times 3 = 32 + 12 = 44$. You'll get the same answer either way, so you can choose the easier method, and when variables are involved, that's usually multiplying first. So

$$3t(5t + 2) = 3t \cdot 5t + 3t \cdot 2 = 15t^2 + 6t.$$

To multiply a single term times an expression, multiply each term of the expression by the multiplier.

$$-3x(5x^2 + 7x - 2) = -3x \cdot 5x^2 - 3x \cdot 7x - 3x \cdot -2$$
$$= -15x^3 - 21x^2 + 6x$$

If you're faced with an expression times an expression, you have to apply the distributive property repeatedly. Treat one expression as the multiplier and distribute to all the terms in the other expression. Then multiply each of the smaller products, using the distributive property again.

$$\underbrace{(x + 5)}_{\text{multiplier}}(x + 3) = \underbrace{(x + 5) \cdot x}_{\text{distribute}} + \underbrace{(x + 5) \cdot 3}_{\text{distribute}}$$
$$= x \cdot x + 5 \cdot x + x \cdot 3 + 5 \cdot 3$$
$$= x^2 + 5x + 3x + 15$$
$$= x^2 + 8x + 15$$

The third reason for parentheses is to show that an expression, not just a term, is being subtracted. If you want to subtract the expression $2x + 9$ from the expression $5x + 12$, and you write $5x + 12 - 2x + 9$, your reader will understand that you want to subtract the $2x$ from the $5x$, but won't know that you want to subtract the 9 as well. To show that you're subtracting the whole $2x + 9$, you need to write $5x + 12 - (2x + 9)$.

To eliminate parentheses that are used to show subtraction of an expression, remember that subtracting is adding the opposite. To subtract $2x + 9$, add $-2x - 9$.

$$5x + 12 - (2x + 9) = 5x + 12 - 2x - 9$$

A subtraction sign in front of parentheses changes the sign of every term in the parentheses.

$$9t - 4 - (2t^2 - 5t + 4) = 9t - 4 - 2t^2 + 5t - 4$$

Combining like terms

Once you've cleared the parentheses, your next step in simplifying the expression is to combine any like terms. **Like terms** are terms with identical variable parts. They must have the same variable (or variables) raised to the same power. The terms $5x$ and $-12x$ are like terms, but $4x^3$ and $3x^2$ are not like, because they don't have the same exponent. The terms $6a^2b^3$ and $9a^3b^2$ are unlike, but $-3a^7b^3$ and $11a^7b^3$ are like, because their variable parts match. Only the multipliers in front, called the **coefficients**, are different.

To combine like terms, add or subtract the coefficients, as indicated. In large expressions, you may want to rewrite the expression with the like terms grouped together first to make the arithmetic easier.

$$\boxed{6x^3} + 4y - 2x^2 - 3xy + 2y^2 - 9x + 8y \boxed{-4x^3} + 5x^2$$
$$= 6x^3 - 4x^3 \boxed{+4y} - 2x^2 - 3xy + 2y^2 - 9x \boxed{+8y} + 5x^2$$
$$= 6x^3 - 4x^3 + 4y + 8y \boxed{-2x^2} - 3xy + 2y^2 - 9x \boxed{+5x^2}$$
$$= \underbrace{6x^3 - 4x^3}_{2x^3} \underbrace{+4y + 8y}_{12y} \underbrace{-2x^2 + 5x^2}_{3x^2} - 3xy + 2y^2 - 9x$$
$$= 2x^3 + 12y + 3x^2 - 3xy + 2y^2 - 9x$$

It's not a requirement, but most people will organize the expression so that terms with the same variable are grouped together, like so:

$$2x^3 + 12y + 3x^2 - 3xy + 2y^2 - 9x = 2x^3 + 3x^2 - 9x - 3xy + 12y + 2y^2$$

EXERCISE 7·3

Simplify each expression by removing parentheses and combining like terms.

1. $3t + 8t$

2. $10x - 6x$

3. $5x + 3y - 2x$

4. $2y - 3 + 5x + 8y - 4x$

5. $6 - 3x + x^2 - 7 + 5x - 3x^2$

6. $(5t + 3) + (t - 12r) - 8 + 9r + (7t - 5)$

7. $(5x^2 - 9x + 7) + (2x^2 + 3x + 12)$

8. $(2x - 7) - (y + 2x) - (3 + 5y) + (8x - 9)$

9. $(3x^2 + 5x - 3) - (x^2 + 3x - 4)$

10. $2y - (3 + 5x) + 8y - (4x - 3)$

Variables on both sides

When variable terms appear on both sides of the equation, add or subtract to eliminate one of them. This should leave a one- or two-step equation for you to solve.

$$3x - 7 = 2x + 4$$
$$_{-2x}_{-2x}$$
$$x - 7 = 4$$
$$x = 11$$

EXERCISE 7·4

Eliminate the extra variable term by adding or subtracting, then solve the equation.

1. $5x - 8 = x + 22$

2. $11x + 18 = 7x - 14$

3. $3x + 18 = 4x - 9$

4. $30 - 4x = 16 + 3x$

5. $11x - 5 = 10 - 4x$

6. $8x - 13 = 12 + 3x$

7. $3x - 5 = 2 - 4x$

8. $1.5x - 7.1 = 8.4 + x$

9. $13 - 9x = 7x - 19$

10. $-5x + 21 = 27 - x$

Simplify before solving

If an equation contains parentheses or has more than two terms on either side, take the time to simplify each side of the equation before you try to solve. If there is a multiplier in front of the parentheses, use the distributive property to multiply and remove the parentheses. Focus on one side at a time and combine like terms. There should be no more than one variable term and one constant term on each side of the equation when you actually start the process of solving by inverse operations.

Get ready to solve: $4(3x - 8) = 10x - (2x + 3) - 1$
Remove parentheses: $12x - 32 = 10x - 2x - 3 - 1$
Combine like terms: $12x - 32 = 8x - 4$

Now you have one variable term and one constant term on each side of the equation. You're ready to actually solve.

Subtract the extra variable term: $12x - 32 = 8x - 4$
$$_{-8x}_{-8x}$$

Add 32 to both sides to eliminate the extra constant term: $4x - 32 = -4$
$$_{+32}_{+32}$$

Divide both sides by 4: $\dfrac{4x}{4} = \dfrac{28}{4}$
$$x = 7$$

If you simplify both sides of an equation, and start solving, only to find that all your variables disappear, that's a sign that your equation either has no solution, or has infinitely many solutions. If the variables disappear and the statement that's left is false, like $0 = 6$, then the equation has no solution. It's a false statement, so no value for the variable will ever make it true. If the variables disappear and the statement is true, like $0 = 0$, then the equation is true for every possible value of the variable. It is solved by any real number. An equation that is true for all real numbers is called an identity.

Simplify the left side and the right side of each equation. Leave no more than one variable term and one constant term on each side. Then solve each equation.

1. $5(x+2)=40$

2. $4(x-7)+6=18$

3. $5(x-4)=7(x-6)$

4. $4(5x+3)+x=6(x+2)$

5. $8(x-4)-16=10(x-7)$

6. $6(2x+9)-30=4(7x-2)$

7. $7(x-1)+2x=12+5(x+1)$

8. $6(x-1)-2x=2(x+1)+4(2-x)$

9. $5(6x+2)+7(4-12x)=35-(6+27x)$

10. $8(2x-5)-2(x-2)=5(x+7)-4(x+8)$

Solving inequalities

When you solve an equation, you find the value of the variable that makes the two sides of the equation identical. Inequalities ask you to find the values that make one side larger than the other. The solution will be a set of numbers, rather than a single value. Your answer needs to give the number that divides solutions from non-solutions, tell which side is which, and whether the divider is a solution or not.

The rules for solving inequalities are the same as the rules for solving equations, with one exception. When you are solving an inequality, and you multiply both sides or divide both sides *by a negative number*, you must switch the direction of the inequality sign. If you're solving $-5x>15$ and you divide both sides by -5, the greater than sign will switch to a less than sign.

$$\frac{-5x}{-5}>\frac{15}{-5}$$
$$x<-3$$

If you multiply both sides of the inequality $\frac{t}{-3}\le 7$ by -3, the less than or equal to sign will reverse to a greater than or equal to sign.

$$\frac{t}{-3}\cdot -3\le 7\cdot -3$$
$$t\ge -21$$

The change only occurs when you multiply or divide by a negative number. If $6x\ge -12$ and you divide by 6, there's no reversal. $x\ge -2$.

Solve each inequality.

1. $3x-5\ge 22$

2. $2x-5>13-4x$

3. $3x+2\le 8x+22$

4. $12x+3<x+36$

5. $t-9\ge 24-10t$

6. $2y-13>4(2-y)$

7. $5x+10(x-1)\ge 95$

8. $5x-4\le 13x+28$

9. $3x-2<2x-3$

10. $-x+5\ge -2+x$

·8· Quadratic equations

When an equation contains a term in which the variable is squared, it's called a **quadratic equation**. Unlike linear equations that have a single solution, quadratic equations generally have two solutions. This may seem odd at first, but if you realize that both 2 and −2, when squared, give you 4, it's easier to see why equations like $x^2 = 4$ have two solutions.

Officially, the standard form of a quadratic equation is $ax^2 + bx + c = 0$, that is, a squared term, a first power term, and a constant all adding to zero, but you'll see lots of rearranged and modified versions. The first power term (also called the x-term) or the constant term may be missing, but if the x^2 term is missing, it's not a quadratic equation. There are several methods for solving quadratic equations. Let's look at the three most common methods.

Square roots

If a quadratic equation contains just a squared term and a constant term, you can solve it by moving the terms to opposite sides of the equal sign and taking the square root of both sides. Remember that there is both a positive and a negative square root of any positive number. That's how you get two solutions. If the constant is not a perfect square, leave solutions in simplest radical form, unless there's a very good reason to use a decimal approximation.

The equation $2x^2 - 64 = 0$ has a squared term, $2x^2$, and a constant term, −64, but no x-term, so it's a good candidate for the square root method.

Move the terms to opposite sides of the equal sign. $\qquad 2x^2 - 64 = 0$

$$2x^2 = 64$$

Divide both sides by 2, to isolate the x^2. $\qquad x^2 = 32$

Take the square root, remembering both the positive and the negative possibility, and put the radical in simplest form. $\qquad x = \pm\sqrt{32} = \pm 4\sqrt{2}$

EXERCISE

8·1

Solve each equation by the square root method.

1. $x^2 = 64$

2. $x^2 - 16 = 0$

3. $x^2 - 8 = 17$

4. $x^2 = 18$

5. $3x^2 = 48$

6. $t^2 - 1{,}000 = 0$

7. $2y^2 - 150 = 0$

8. $9x^2 = 4$

9. $64y^2 = 25$

10. $4x^2 - 15 = 93$

Factoring

The square root method is quick and easy, but it only works for equations that have only a squared term and a constant term. If the equation has an x-term, the square root method won't be any help. That's the bad news. The good news is that many quadratic equations that do have an x-term can be solved quickly by a method commonly called factoring. Officially, it's an application of the **zero product property**. That's the name for the observation that if you think of two numbers that multiply to zero, one of them has to be zero. There's no way you can get zero by multiplying unless at least one of the factors is zero.

If the product of two factors is zero, then at least one of the factors is zero, so if a quadratic can be written as the product of two factors, you know that one of those factors equals zero. By looking at each one to see what value of the variable would make it zero, you can break the quadratic equation down into two simple equations.

Multiplying

Before you start trying to work backward, let's look at the kinds of multiplication that can create a quadratic. There are two basic types. If you multiply a variable, or a single variable term, times the sum (or difference) of a variable term and a constant, for example, $3x(2x + 5)$, you'll get a squared term and an x-term. $3x(2x + 5) = 6x^2 + 15x$. This is the easier of the two types. It's easier to multiply because you just use the distributive property once, and it's easier to undo.

The second type of multiplication that produces a quadratic is a pattern called FOIL and this one produces all three terms. It's the method of multiplication you use if you're faced with an expression with two terms times another expression with two terms. To multiply, you have to apply the distributive property repeatedly. The name FOIL comes from a shortcut.

Suppose you need to multiply $(x + 5)(x + 3)$. Treat the $(x + 5)$ as the multiplier and distribute it to both the terms in $(x + 3)$.

$$\underbrace{(x + 5)}_{multiplier}(x + 3) = (x + 5) \cdot x + (x + 5) \cdot 3$$

Then multiply $(x + 5) \cdot x$ and $(x + 5) \cdot 3$ by using the distributive property again.

$$\underbrace{(x + 5)}_{multiplier}(x + 3) = \underbrace{(x + 5) \cdot x}_{distribute} + \underbrace{(x + 5) \cdot 3}_{distribute}$$
$$= x \cdot x + 5 \cdot x + x \cdot 3 + 5 \cdot 3$$
$$= x^2 + 5x + 3x + 15$$

You end up with four terms, although you may have already noticed that two of them are like terms and can be combined. The letters in the word FOIL are a reminder of how to get those four terms. The letters in FOIL stand for First, Outer, Inner, Last:

First $(x+5)(x+3) \Rightarrow x \cdot x$

Outer $(x+5)(x+3) \Rightarrow x \cdot 3$

Inner $(x+5)(x+3) \Rightarrow 5 \cdot x$

Last $(x+5)(x+3) \Rightarrow 5 \cdot 3$

You will often find that there are like terms that can be combined after the four multiplications are performed.

$$(2x-3)(5x+7) = (2x)(5x)+(2x)(7)+(-3)(5x)+(-3)(7)$$
$$= 10x^2 +14x-15x-21$$
$$= 10x^2 -x-21$$

EXERCISE

8·2

Use the FOIL rule to multiply. Give your answers in simplest form.

1. $(x+8)(x+2)$

2. $(y-4)(y-9)$

3. $(t-2)(t+6)$

4. $(2x+8)(x-3)$

5. $(y-9)(3y+1)$

6. $(5x-6)(3x+4)$

7. $(6x-1)(x+5)$

8. $(1-3b)(5+2b)$

9. $(3x-7)(2x+5)$

10. $(5-2x)(5x-2)$

Factoring $ax^2 + bx$

An expression like $3x^2 + 9x$, a squared term and a first power term, is the result of multiplying a single term times a binomial. To factor it, realize this is the result of using the distributive property, and think about what the problem might have been. What do the two terms have in common? Both have a coefficient that is divisible by 3 and both have an x. Use $3x$ as the multiplier. $3x^2 + 9x = 3x$(some binomial). Divide each term of $3x^2 + 9x$ by $3x$ to find that binomial. $3x^2 + 9x = 3x(x + 3)$. Check by multiplying.

Factoring $x^2 + bx + c$

An expression of the form $x^2 + bx + c$ can often be factored to the product of two expressions, each with two terms. (There are some that can't be factored, but there's a way around that coming up later.) If you can factor $x^2 + bx + c$, it will factor to $(x+r)(x+t)$. The product of r and t will equal the constant term, c, and the sum or difference of r and t will produce the middle coefficient, b. Let's look at an example.

To factor $x^2 + 5x + 6$, you first need to find a pair of numbers that multiply to 6. Factors of 6 are 1 and 6 or 2 and 3, so you have two possibilities. Then you have to choose the pair that add to 5, so you want 2 and 3. Put those in the places of r and t and you $x^2 + 5x + 6 = (x+3)(x+2)$. You can check your factors by multiplying, using the FOIL rule.

You may have noticed that 6 and 1, the other possible factor pair, subtract to 5. Why can't you use those? It's all about the signs. When the constant is positive, the two numbers must have the same sign, and when you add numbers with the same sign, you add and keep the sign. When the constant is negative, one of the numbers gets a positive sign and one a negative sign, and adding numbers with opposite signs looks like subtracting.

Here's a summary of what can happen. When you look for the factors of the constant, don't worry about the signs. Make a list of the possible factor pairs for your constant. Check the chart to see if you need to add or subtract to the middle term, and then follow the chart for signs.

$x^2 + bx + c$	Add to b	$(x+r)(x+t)$	Both get the sign of b	$x^2 + 5x + 6 = (x+3)(x+2)$
$x^2 - bx + c$		$(x-r)(x-t)$		$x^2 - 5x + 6 = (x-3)(x-2)$
$x^2 + bx - c$	Subtract to b	$(x+r)(x-t)$	Bigger number gets the sign of b	$x^2 + 5x - 6 = (x+6)(x-1)$
$x^2 - bx - c$		$(x-r)(x+t)$		$x^2 - 5x - 6 = (x-6)(x+1)$

EXERCISE 8·3

Factor each expression. Check your answers by multiplying.

1. $x^2 + 12x + 35$

2. $x^2 + 11x + 28$

3. $x^2 - 8x + 15$

4. $x^2 - 7x + 12$

5. $x^2 + x - 20$

6. $x^2 - 2x - 3$

7. $x^2 - 11x + 18$

8. $x^2 - 9x - 22$

9. $x^2 + 10x - 39$

10. $x^2 + 12x + 32$

Solving by factoring

To solve by factoring, first, put the equation in standard form: $x^2 + bx + c = 0$.

$$5x = 4 - 6x^2$$
$$6x^2 + 5x - 4 = 0$$

With all terms on one side of the equation equal to zero, factor the quadratic expression.

$$6x^2 + 5x - 4 = 0$$
$$(2x - 1)(3x + 4) = 0$$

Set each factor equal to zero.

$$(2x - 1)(3x + 4) = 0$$
$$2x - 1 = 0 \qquad 3x + 4 = 0$$

Finally, solve the resulting equations.

$$2x - 1 = 0 \qquad 3x + 4 = 0$$
$$2x = 1 \qquad 3x = -4$$
$$x = \frac{1}{2} \qquad x = \frac{-4}{3}$$

If your quadratic equation has a squared term and an x-term, but no constant term, such as $6x^2 + 15x = 0$, you can factor it as x times something, for example, $x(6x + 15) = 0$. You could also write it as $3x(2x + 5) = 0$, but either version will get you to the correct answer.

EXERCISE

8·4

Solve each equation by factoring.

1. $x^2 + 5x + 6 = 0$

2. $x^2 + 12 = 7x$

3. $y^2 + 3y = 8 + y$

4. $a^2 - 3a - 4 = 6$

5. $20 = x^2 + x$

6. $x^2 + 5 = 6x$

7. $x^2 + 3x = 0$

8. $x^2 = 5x$

9. $2x^2 - 1 = 3x^2 - 2x$

10. $2x^2 - x = 0$

Quadratic formula

The **quadratic formula** is a shortcut to the solution of a quadratic equation. It's a bit complicated, but it will always work. It doesn't matter if you have an x-term or not, if you have a constant or not, whether the quadratic is factorable or not. The quadratic formula will give you a solution (or tell you clearly that there is none).

The quadratic formula says that if $ax^2 + bx + c = 0$, then $x = \dfrac{-b \pm \sqrt{b^2 - 4ac}}{2a}$. You just pick the values of a, b, and c out of the equation, plug them into the formula, and simplify. Be certain your equation is in $ax^2 + bx + c = 0$ form before deciding on the values of a, b, and c.

To solve $5x = 2 - 3x^2$, first put the equation in standard form. Compare $3x^2 + 5x - 2 = 0$ to $ax^2 + bx + c = 0$, and you find that $a = 3$, $b = 5$, and $c = -2$. Plug those values into the formula.

$$x = \frac{-b \pm \sqrt{b^2 - 4ac}}{2a}$$

$$x = \frac{-5 \pm \sqrt{5^2 - 4 \cdot 3 \cdot (-2)}}{2 \cdot 3}$$

Follow the order of operations and watch your signs as you simplify.

$$x = \frac{-5 \pm \sqrt{25 + 24}}{6}$$

$$x = \frac{-5 \pm \sqrt{49}}{6}$$

$$x = \frac{-5 \pm 7}{6}$$

The two solutions that are typical of quadratic equations come from that plus or minus sign.

$$x = \frac{-5 \pm 7}{6}$$

$$x = \frac{-5 + 7}{6} \qquad x = \frac{-5 - 7}{6}$$

$$x = \frac{2}{6} \qquad x = \frac{-12}{6}$$

$$x = \frac{1}{3} \qquad x = -2$$

EXERCISE 8·5

Solve each equation by the quadratic formula. If necessary, leave answers in simplest radical form.

1. $x^2 + 4x - 21 = 0$

2. $t^2 = 10 - 3t$

3. $y^2 - 4y = 32$

4. $x^2 = 6 + x$

5. $6x + x^2 = 9$

6. $t^2 + 6t - 15 = 0$

7. $4x^2 - 3 = x$

8. $3x^2 - 1 = 2x$

9. $x + 5 = 3x^2 - x$

10. $6x^2 - 2 = x$

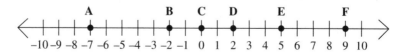

·9· Segments and angles

The building blocks of geometry are points, lines, and planes. Such fundamental ideas are sometimes difficult or impossible to define, and since you have to start somewhere, we take those terms as undefined, and then we start to combine them into more complex figures.

A point is named with a single upper case letter, for example, point P. A line is named by two points with a double-pointed arrow over the top.

Segments, midpoints, and bisectors

A **segment** is a portion of a line between two endpoints. A segment is named by its endpoints with a rule drawn over the top. The segment with endpoints A and B can be named as $\overline{AB}$ or $\overline{BA}$.

A **line** goes on forever, so it is not possible to measure a line, to put a number on it, because it has no beginning and no end. Segments, because they have two endpoints, have a finite length. A segment can be assigned a number that tells how long it is. What number describes the length of a segment will depend on what units of measurement (inches, centimeters, etc.) are used.

When you use a ruler to measure the length of a segment, you may align the zero of the ruler with one end of the segment and see where the other end falls, but you don't have to start at zero. You can place the ruler next to the segment and note where each endpoint falls, and then subtract. The number that corresponds to an endpoint is called its **coordinate**. In Figure 9-1, the coordinate of point D is 2. Point A has a coordinate of -7. The distance between two points is the absolute value of the difference of their coordinates, or the number you get by subtracting the smaller coordinate from the larger one. In this figure the distance between point B and point E is $|-2-5| = 7$ units. (Remember that $|\ |$ indicates absolute value.) Distance will never be negative, regardless of direction of the line.

Figure 9-1

The length of a segment is the distance between its endpoints, so the length of $\overline{EF}$ is the distance between E and F, which is $|5-9| = 4$ units. The length of $\overline{AB}$ is $|-7--2| = 5$. Write AB, without any symbol over the top, to mean the length of segment $\overline{AB}$. The length of $\overline{BE}$ is $BE = 7$ units, and $AB = 5$ units is the length of $\overline{AB}$. Two segments that have the same length are congruent. If $AC = DF$, then $\overline{AC} \cong \overline{DF}$.

The **midpoint** of a segment is a point on the segment that is the same distance from one endpoint as it is from the other. Point M is the midpoint of $\overline{AB}$ if M is a point on $\overleftrightarrow{AB}$ and $AM = MB$. Each line segment has only one midpoint. A **bisector** of a segment is any line or segment that passes through the midpoint. A segment has exactly one midpoint, but it can have many bisectors. If the bisector of a segment also makes a right angle with the segment, it is called the **perpendicular bisector**. Every point on the perpendicular bisector of a segment is equidistant from the endpoints of the segment.

Use the figure to answer questions 1 through 15. In questions 1 through 7, find the length of each segment.

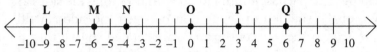

1. $\overline{LP}$

2. $\overline{MO}$

3. $\overline{NQ}$

4. $\overline{NM}$

5. $\overline{QL}$

6. $\overline{MP}$

7. $\overline{ON}$

8. Name a segment that is congruent to $\overline{LM}$.

9. Name a segment that is congruent to $\overline{MO}$.

10. Name a segment that is congruent to $\overline{MP}$.

11. What is the midpoint of $\overline{OQ}$?

12. If A is the midpoint of $\overline{NO}$, what is the coordinate of A?

13. If B is the midpoint of $\overline{MN}$, what is the coordinate of B?

14. If M is the midpoint of $\overline{LC}$, what is the coordinate of C?

15. If P is the midpoint of $\overline{ND}$, what is the coordinate of D?

Angles and bisectors

An **angle** is the figure formed when portions of two lines called rays shoot out from a common endpoint. The common endpoint is called the **vertex**, and the line parts are the **sides** of the angle. In Figure 9-2, the point Y is the vertex of the angle, and $\overrightarrow{YX}$ and $\overrightarrow{YZ}$ are the sides. Because the sides go on through X and Z, but not beyond Y, the symbol over the top only has one arrowhead.

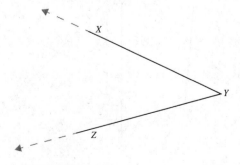

Figure 9-2

Angles are named by naming a point on one side, then the vertex, then a point on the other side. The three letters outline the angle. The angle in Figure 9-2 can be named ∠*XYZ* or ∠*ZYX*. The vertex letter must be in the middle.

If there is only one angle with a particular vertex and no confusion is possible, the angle can be named by just its vertex letter. The angle in Figure 9-2 can be named ∠*Y*, because it is the only angle with vertex *Y*, but in Figure 9-3, ∠*P* is not a useful name because there are three angles with vertex *P*: ∠*RPS*, ∠*QPS*, and ∠*RPQ*.

When there are several different angles at the same vertex, it can be helpful to number the angles. ∠*RPQ* can be named as ∠1 and ∠*QPS* as ∠2.

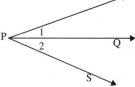

Figure 9-3

Measuring angles

The measurement of an angle has nothing to do with how long the sides are. The size of the angle is a measurement of the rotation or separation between the sides. A full rotation, all the way around a circle, is 360°. If the sides of the angle are rotated so that they point in opposite directions, the measure of the angle is 180°. The hands of a clock make a 180° angle at 6:00. At 6:15, they form only a quarter rotation, a 90° angle.

The instrument used for measuring angles is called a **protractor**. It is a semicircle with markings for the degrees from 0 to 180. Most protractors have two scales: one goes clockwise and one counterclockwise. Be certain you read the coordinates for both sides of the angle on the same scale.

If a protractor is placed on an angle that has its vertex at point *Y*, you can look at where the sides cross the protractor's scale to determine the measurement of the angle. If the coordinate of side *XY* is 140 and the coordinate of side *YZ* is 50, the measure of ∠*ZYX* = 140 − 50 = 90°. A small *m*, written before the name of the angle, denotes the measure of the angle: *m*∠*ZYX* = 90°.

If two angles have the same measurement, then they are congruent. ∠*XYZ* ≅ ∠*ABC* means that the measurement of ∠*XYZ* is equal to the measurement of ∠*ABC*.

Bisectors

The **bisector** of an angle is a line or segment that passes through the vertex of the angle and divides the angle into two angles of equal size. Line segment $\overline{AB}$ is the bisector of ∠*CAD* if ∠*CAB* = ∠*BAD*. An angle has exactly one bisector.

EXERCISE

9·2

Use the figure to answer questions 1 through 10.

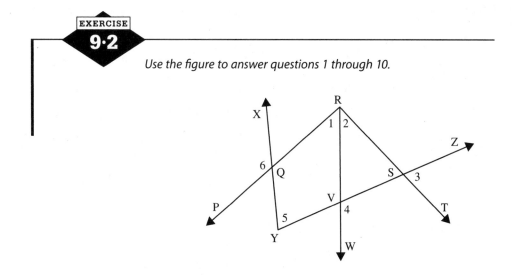

1. Name ∠1 with three letters.

2. Name ∠4 with three letters.

3. Name three angles with a vertex at R.

4. Give another name for ∠6.

5. Give another name for ∠TSZ.

6. What is the vertex of ∠5?

7. Name the sides of ∠2.

8. True or False: ∠WVS is another name for ∠WVZ.

9. True or False: ∠YQV is another name for ∠5.

10. True or False: ∠RVY is another name for ∠4.

Use the following figure to find the measure of each angle in questions 11 through 20.

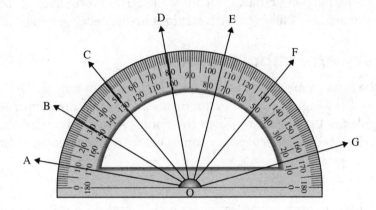

11. ∠AOF

12. ∠BOE

13. ∠COD

14. ∠DOB

15. ∠EOG

16. ∠FOB

17. ∠GOC

18. ∠EOA

19. ∠COB

20. ∠DOG

Use the figure to answer questions 21 through 25.

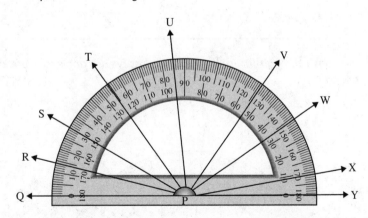

21. True or False: Since m∠TPU = m∠WPX, $\overrightarrow{PU}$ is the bisector of ∠TPX.

22. True or False: $\overrightarrow{PT}$ is the bisector of ∠SPU.

23. Name the bisector of ∠QPS.

24. Where does the bisector of ∠TPW fall on the scale of the protractor?

25. Where does the bisector of ∠RPU fall on the scale of the protractor?

Angle relationships

Once you know how to measure an angle, your next step is to look at different ways to classify angles and talk about their relationships.

Acute, right, obtuse

Angles can be classified according to their measure. Angles that measure less than 90° are called **acute** angles. An angle that measures exactly 90° is a **right** angle. Two lines (or segments) that form a right angle are **perpendicular.** If an angle measures more than 90° but less than 180°, the angle is **obtuse**. An angle of 180° is called a **straight** angle, because it looks like a straight line.

Complementary and supplementary

Two angles whose measurements add to 90° are complementary angles. Each angle is the **complement** of the other. Two angles whose measurements add to 180° are **supplementary** angles. Each angle is the **supplement** of the other. Keep the names straight by remembering that 90 comes before 180 numerically and *complementary* comes before *supplementary* alphabetically.

Linear pairs

Two angles that have a common vertex, share a side, and do not overlap are **adjacent angles**. When adjacent angles have exterior sides that form a line (or a straight angle) they are called a **linear pair**. The two angles in a linear pair are always supplementary.

Vertical angles

When two lines intersect, the **X**-shaped figure they create has two pairs of vertical angles. **Vertical angles** have a common vertex, but no shared sides. The two angles in a pair of vertical angles are always congruent.

For questions 1 through 10, use the figure to classify each angle as acute, right, obtuse, or straight.

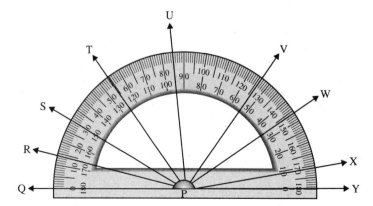

1. ∠*RPS*

2. ∠*QPY*

3. ∠*UPW*

4. ∠*TPW*

5. ∠*QPR*

6. ∠*UPX*

7. ∠*RPV*

8. ∠*QPX*

9. ∠*WPX*

10. ∠*XPR*

11. Find the measure of the complement of an angle of 53°.

12. Find the measure of the complement of an angle of 31°.

13. Find the measure of the supplement of an angle of 47°.

14. Find the measure of the supplement of an angle of 101°.

Use the next figure to answer questions 15 through 20.

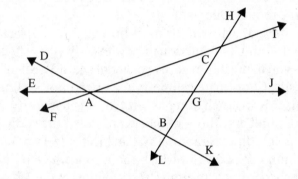

15. Name an angle that forms a pair of vertical angles with ∠*HCI*.

16. Name an angle that forms a linear pair with ∠*EAD*.

17. Name an angle that forms a pair of vertical angles with ∠*GBA*.

18. Name an angle that forms a linear pair with ∠*GBK*.

19. Name an angle that is congruent to ∠*BAG*.

20. Name an angle that is supplementary to ∠*JGC*.

Coordinate geometry

Arithmetic works with numbers, algebra communicates through variables and other symbols, and geometry uses figures and pictures. The place where they all meet is a branch of mathematics known as coordinate geometry, or simply, graphing. It's a system that takes a flat surface, called a plane, and gives you a way to identify every spot on that plane. Once you can do that, you can use points in the plane to illustrate patterns.

If a friend asks you to meet him on Main Street, you have a place to start looking for him, but you don't know exactly where he'll be. On the other hand, if he asks you to meet him at the corner of Main Street and River Road, you can find him easily. Coordinate geometry uses two number lines, one horizontal and one vertical, crossing each other at zero, to let you identify any location in the plane by a pair of numbers. Those numbers are called the coordinates of the point.

The coordinate graphing system locates every point in the plane by an ordered pair of numbers, (x, y), in which the x-coordinate indicates horizontal movement and the y-coordinate indicates vertical movement. When talking about two different points, each of which has coordinates (x, y), you can use a small subscript to distinguish between them. The first point is (x_1, y_1) and the second is (x_2, y_2). The horizontal, or x-axis, and the vertical, or y-axis, divide the plane into four quadrants. The point $(4, 7)$, for example, is 4 units to the right and 7 units up from a starting point called the origin. This point is graphed in Figure 10-1.

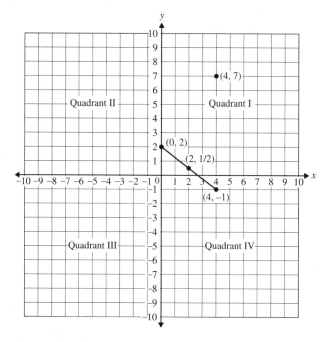

Figure 10-1

Distance

The distance between two points can be calculated by means of the distance formula: $d = \sqrt{(x_2 - x_1)^2 + (y_2 - y_1)^2}$. The formula is an application of the Pythagorean theorem, in which the distance between (x_1, y_1) and (x_2, y_2) is the hypotenuse of the right triangle. For example, Figure 10-1 shows the points (4, –1) and (0, 2) connected by a line segment. The distance between the points (4, –1) and (0, 2), can be calculated like this:

$$d = \sqrt{(4-0)^2 + (-1-2)^2} = \sqrt{16+9} = 5$$

If the two points fall on a vertical line or on a horizontal line, the distance will simply be the difference in the coordinates that don't match. The distance between (–3, 2) and (–3, 6) is $6 - 2 = 4$ units. The distance between (5, 7) and (–8, 7) is $5 - -8 = 5 + 8 = 13$ units.

Midpoint

The **midpoint** of the segment that connects (x_1, y_1) and (x_2, y_2) can be found by averaging the x-coordinates and averaging the y-coordinates.

$$M = \left(\frac{x_1 + x_2}{2}, \frac{y_1 + y_2}{2} \right)$$

The midpoint of the segment connecting (4, –1) and (0, 2) is

$$M = \left(\frac{4+0}{2}, \frac{-1+2}{2} \right) = \left(2, \frac{1}{2} \right)$$

EXERCISE 10·1

For questions 1 through 5, find the distance between the given points by using the distance formula.

1. (4, 5) and (7, 1)

2. (6, 2) and (15, 14)

3. (–7, –1) and (5, 4)

4. (5, 3) and (8, 6)

5. (–4, 2) and (3, 2)

For questions 6 through 10, find the midpoint of the segment with the given endpoints by using the midpoint formula.

6. (2, 3) and (6, 9)

7. (–3, 1) and (–1, 7)

8. (–5, –3) and (–1, –1)

9. (8, 0) and (0, 8)

10. (0, –2) and (4, –4)

Slope

The **slope** of a line is a measurement of the rate at which it rises or falls. A rising line has a positive slope and a falling line has a negative slope. A horizontal line has a slope of zero, and the farther from zero the slope is, the steeper the line is.

Suppose you choose two points on the line and move from one point to the other by going up or down and then across. The amount you move up or down is called the **rise**. It's a positive number if you move up and a negative number if you move down. You can find the rise by subtracting the y-coordinates of the points. The **run** is the amount you moved left or right. It's positive if you move right and negative if you move left. You can find the rise by subtracting the x-coordinates. Once you know the rise and the run, you can find the slope, which is usually denoted by the letter m.

$$m = \frac{\text{rise}}{\text{run}} = \frac{y_2 - y_1}{x_2 - x_1}$$

The slope of the line through the points (4, –1) and (0, 2) is

$$m = \frac{2 - -1}{0 - 4} = -\frac{3}{4}$$

A vertical line rises but never runs. It goes up and down, but not left or right. We say that a vertical line has no slope or that its slope is undefined.

Equation of a line

The equation of a line includes two pieces of information: the slope of the line and the point where it crosses the y-axis, called the **y-intercept**. Those two numbers give you all you need to draw the graph of the equation, a picture of all the possible solutions. Traditionally, the letter m stands for the slope and the letter b stands for the y-intercept. In that case, the equation of a line $y = mx + b$.

If you know the slope and the y-intercept, you can just pop them into their places, and you have the equation of the line. If a line has a slope of -2 and a y-intercept of 7, its equation is $y = -2x + 7$. If you have the equation of a line, and it's arranged in $y = mx + b$ form, you can read the slope and y-intercept. The equation $y = \frac{4}{5}x - 3$ has a slope of $\frac{4}{5}$, and crosses the y-axis at (0, −3). Being able to pick out the slope and y-intercept will be helpful when you need to draw the graph, but be certain your equation is in $y = mx + b$ form before you try to identify the slope and y-intercept. Any rearrangement could deceive you.

You don't always have the slope and y-intercept, of course. Sometimes you have the slope and a point that's not the y-intercept. Other times you have two points. You can still find the equation of the line by using **point-slope form**: $y - y_1 = m(x - x_1)$.

If you know the slope and a point, you plug the slope in for m, the x-coordinate of the point in for x_1, and the y-coordinate of the point in for y_1 in the point-slope form. If a line has a slope of $\frac{2}{3}$ and passes through the point (6, −5), then the equation of the line is $y + 5 = \frac{2}{3}(x - 6)$. Of course, usually you'd simplify that by clearing the parentheses and isolating y.

$$y + 5 = \frac{2}{3}(x - 6)$$

$$y + 5 = \frac{2}{3}x - 4$$

$$y = \frac{2}{3}x - 9$$

If you know two points on the line, you'll need to first use the slope formula to find the slope, and then use the slope and one of your points—either one—to plug into point-slope form. The line that passes through the points $(-3, -1)$ and $(4, 6)$ has a slope of

$$m = \frac{y_2 - y_1}{x_2 - x_1} = \frac{-1-6}{-3-4} = \frac{-7}{-7} = 1$$

Using the slope of 1 and the point $(4, 6)$, the equation of the line is

$$y - y_1 = m(x - x_1)$$
$$y - 6 = 1(x - 4)$$
$$y = x + 2$$

EXERCISE 10·2

For questions 1 through 5, find the slope of the line that passes through the two points given.

1. $(-5, 5)$ and $(5, -1)$

2. $(6, -4)$ and $(9, -6)$

3. $(3, 4)$ and $(8, 4)$

4. $(4, 6)$ and $(8, 7)$

5. $(7, 2)$ and $(6, 5)$

For questions 6 through 10, identify the slope and y-intercept of each line.

6. $y = 9x + 4$

7. $y = -3x + 2$

8. $y + 5 = \frac{1}{2}x$

9. $x + y = 6$

10. $2x - 3y = 12$

For questions 11 through 20, find the equation of the line described.

11. Slope of -2 and y-intercept of 7.

12. Slope of $\frac{1}{2}$ and y-intercept of -9.

13. Slope of $-\frac{1}{5}$ and a y-intercept of $\frac{2}{3}$.

14. Slope of 5 passing through the point $(3, -1)$.

15. Slope of -3 and passing through the point $(1, 1)$.

16. Slope of 4 and passing through the point $(4, 13)$.

17. Slope of 2 and passing through the point $(8, 0)$.

18. Slope of $\frac{3}{4}$ and passing through the point $(-4, 5)$.

19. Passing through the points $(4, -5)$ and $(6, 3)$.

20. Passing through the points $(-1, -1)$ and $(5, 11)$.

Graphing linear equations

The graph of a linear equation is a picture of all the possible pairs of numbers (x, y) that would solve the equation. Every pair of numbers that makes a true statement when plugged in is a point on the line. The basic way to graph any equation is to look for points that solve the equation, but for linear equations, there are some shortcuts as well.

Table of values

The most straightforward way to graph an equation is to choose several values for x, substitute each value into the equation, and calculate the corresponding values for y. This information can be organized into a table of values. Geometry tells us that two points determine a line, but when building a table of values, it is wise to include several more, so that any errors in arithmetic will stand out as deviations from the pattern.

For the equation $2x + y = 8$, you could make a table of values by choosing several values for x, plugging each one into the equation, and calculating the corresponding value of y. You can see this line on the graph in Figure 10-2.

x	$2x + y = 8$	y	(x, y)
-3	$2(-3) + y = 8$ $-6 + y = 8$	14	$(-3, 14)$
-2	$2(-2) + y = 8$ $-4 + y = 8$	12	$(-2, 12)$
-1	$2(-1) + y = 8$ $-2 + y = 8$	10	$(-1, 10)$
0	$2(0) + y = 8$ $0 + y = 8$	8	$(0, 8)$
1	$2(1) + y = 8$ $2 + y = 8$	6	$(1, 6)$
2	$2(2) + y = 8$ $4 + y = 8$	4	$(2, 4)$
3	$2(3) + y = 8$ $6 + y = 8$	2	$(3, 2)$

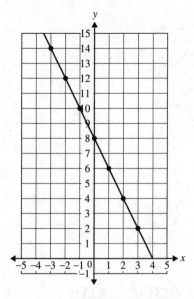

Figure 10-2

Slope and *y*-intercept

To draw the graph of a linear equation quickly, put the equation in slope-intercept or $y = mx + b$ form. The value of b is the y-intercept of the line and the value of m is the slope of the line. Begin by drawing the x- and y-axis on graph paper, and plotting the y-intercept, a point on the y-axis, then count the rise and run and plot another point. Repeat a few times and connect the points to form a line. In fact, you can draw the line this way if you know the slope and any point on the line, but the y-intercept is easy to spot in the equation.

To graph $y = \dfrac{3}{5}x - 2$, first identify the slope of $\dfrac{3}{5}$ and the y-intercept of -2. Plot a point at the y-intercept $(0, -2)$. From that point, count up 3 (along the y-axis) and 5 to the right (along the x-axis), and place another point at $(5, 1)$. From $(5, 1)$ count up 3 and 5 to the right again, and plot another point at $(10, 4)$. Connect the points and extend the line. (See Figure 10-3.)

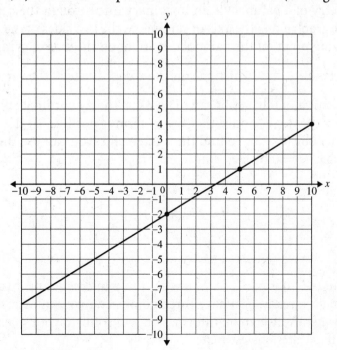

Figure 10-3

Draw the graph of each of the following equations.

1. $y = -\dfrac{3}{4}x + 1$

2. $2x - 3y = 9$

3. $y = -4x + 6$

4. $6x + 2y = 18$

5. $x - 2y = 8$

6. $y = -3x - 4$

7. $y - 6 = 3x + 1$

8. $3x + 5y = 15$

9. $2y = 5x - 6$

10. $3x - 2y - 6 = 0$

Graphs of quadratic equations

Graphing an equation of the form $y = ax^2 + bx + c$ produces a cup-shaped graph called a **parabola**. Some information about the graph can be gathered from the equation without much effort and can help you construct a table of values and plot the graph. If a is a positive number, the parabola opens upward, like a cup that could hold water, but if a is negative, the parabola turns down, and all the water spills out. (See Figure 10-4.) The solutions of the equation $ax^2 + bx + c = 0$ are the x-intercepts, the points where the graph crosses the x-axis, and the constant term is the y-intercept $(0, c)$.

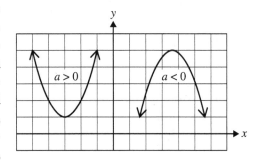

Figure 10-4

Finding axis of symmetry and vertex

The **axis of symmetry** of a parabola is an imaginary line through the center of the parabola. If you were to fold the graph along the axis of symmetry, the two sides of the parabola would match. The **vertex**, or turning point, of the parabola sits right on the axis of symmetry. The equation of the axis of symmetry is $x = \dfrac{-b}{2a}$, which means the x-coordinate of the vertex is $\dfrac{-b}{2a}$. Plugging that x-value into the equation of the parabola will give you the y-coordinate of the vertex.

To graph $y = x^2 + 4x - 5$, notice that the x^2 term is positive, so the parabola will open up, and its y-intercept will be $(0, -5)$. To find the x-intercepts, you can use the quadratic formula, or you can solve $x^2 + 4x - 5 = 0$ by factoring.

$$x^2 + 4x - 5 = 0$$
$$(x + 5)(x - 1) = 0$$
$$x + 5 = 0 \qquad x - 1 = 0$$
$$x = -5 \qquad x = 1$$

The x-intercepts are $(-5, 0)$ and $(1, 0)$. Knowing the y-intercept and two x-intercepts, you can almost draw the graph already. You have three points, and you know it should be shaped like a cup that can hold water. A little more information will confirm your guess, however. The dotted line in Figure 10-5 illustrates one estimate.

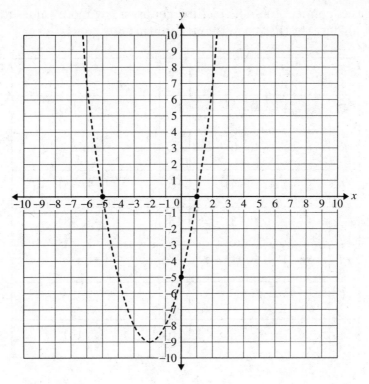

Figure 10-5

To confirm this estimate, we need to find the vertex. Since the axis of symmetry for $y = x^2 + 4x - 5$ is $x = \dfrac{-b}{2a} = \dfrac{-4}{2 \cdot 1} = -2$, the parabola will be symmetric across the vertical line $x = -2$. Plug -2 into $y = x^2 + 4x - 5$ and you find that the y-coordinate of the vertex is $y = (-2)^2 + 4 \cdot -2 - 5 = 4 - 8 - 5 = -9$. The vertex is $(-2, -9)$. This tells you where the bottom of the cup is. (See Figure 10-6.)

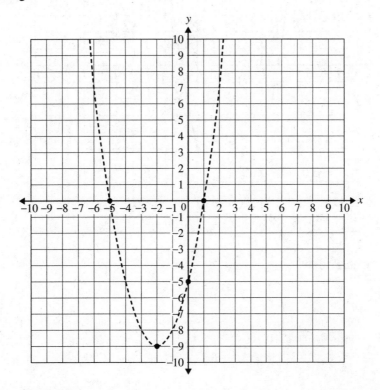

Figure 10-6

If you want an easier connect-the-dots picture, choose a few more values for *x* and make a table of values. Figure 10-7 graphs these points and confirms our guess.

x	$x^2 + 4x - 5$	y	(x, y)
-7	$(-7)^2 + 4(-7) - 5$	16	$(-7, 16)$
-6	$(-6)^2 + 4(-6) - 5$	7	$(-6, 7)$
-4	$(-4)^2 + 4(-4) - 5$	-5	$(-4, -5)$
-3	$(-3)^2 + 4(-3) - 5$	-8	$(-3, -8)$
-1	$(-1)^2 + 4(-1) - 5$	-8	$(-1, -8)$
2	$2^2 + 4 \cdot 2 - 5$	7	$(2, 7)$
3	$3^2 + 4 \cdot 3 - 5$	16	$(3, 16)$

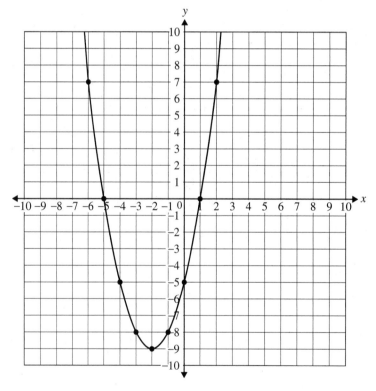

Figure 10-7

For questions 1 through 5, find the x- and y-intercepts of each parabola.

1. $y = x^2 - 4x + 3$

2. $y = x^2 - 4x - 5$

3. $y = x^2 + 2x$

4. $y = x^2 - 7x + 12$

5. $y = x^2 - 2x + 1$

For questions 6 through 10, find the vertex of each parabola.

6. $y = x^2 - 8x + 15$

7. $y = x^2 + 4x - 2$

8. $y = 2x^2 - 4x + 3$

9. $y = -x^2 + 6x - 7$

10. $y = -x^2 + 4x + 7$

For questions 11 through 20, graph each quadratic equation. Use the vertex and intercepts to help make a table of values.

11. $y = 2x^2 - 1$

12. $y = -x^2 + 8x$

13. $y = x^2 + 2x - 15$

14. $y = x^2 - 6x + 1$

15. $y = x^2 - 4x + 3$

16. $y = -x^2 + 2x + 5$

17. $y = x^2 + 6x + 9$

18. $y = 4 - x^2$

19. $y = 2x^2 + 4x - 1$

20. $y = x^2 - 9$

Polygons

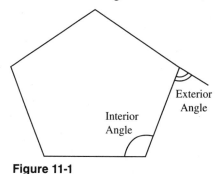

Ask a friend to draw a picture of a building and he'll probably start with a four-sided picture. Not all buildings have four sides, of course. The Pentagon, in Washington, DC, is one obvious example, but a building on an oddly shaped piece of land might have only three sides, like the Flatiron Building in New York City. Even when the building has four sides, parts of it may have other shapes. If you've built a birdhouse or a gingerbread house, you probably needed some five-sided pieces to support those sloping roofs. Understanding all those different shapes, called polygons, is an important part of geometry.

A **polygon** is a closed figure formed by three or more line segments that intersect only at their endpoints. The line segments are the sides of the polygon, and the points where the sides meet are the vertices of the polygon. If all the sides of a polygon are congruent (that is, have equal measure), the polygon is **equilateral**.

At each vertex, the two sides meet to form an interior angle. If all the interior angles of the polygon have the same measure, the polygon is **equiangular**. If one of the sides is extended, an exterior angle is created. (See Figure 11-1.)

Figure 11-1

A polygon that is both equilateral and equiangular is a regular polygon. When a triangle is equilateral, it is automatically equiangular, but that is not the case with other polygons. An equilateral triangle is regular, because all three of its sides are the same length and all its angles measure 60°. A square is a regular polygon because all its angles are right angles and all its sides are congruent.

Polygons are given specific names according to the number of their sides. Common polygons include:

- ◆ 3 sides: triangle
- ◆ 4 sides: quadrilateral
- ◆ 5 sides: pentagon
- ◆ 6 sides: hexagon
- ◆ 8 sides: octagon
- ◆ 10 sides: decagon

Triangles

Triangles are named by the three letters on their vertices or corners, and those three letters are usually preceded by a small triangle symbol. A triangle with vertices X, Y, and Z is named $\triangle XYZ$.

Triangles are classified either by the size of their angles or by the number of congruent sides they have. All triangles have three angles, and those three angles always add up to 180°. If the triangle contains a right angle, it's a **right triangle**, and if it contains an obtuse angle, it's an **obtuse triangle**. An **acute triangle** has three acute angles.

A triangle with three congruent sides is an equilateral triangle, and one with two congruent sides is **isosceles**. If all sides are different lengths, the triangle is **scalene**.

$\triangle XYZ$ has the following measurements:

side $\overline{XY}$ = 8 cm	side $\overline{YZ}$ = 12 cm	side $\overline{XZ}$ = 8 cm
$m\angle X = 98°$	$m\angle Y = 41°$	$m\angle Z = 41°$

$\triangle XYZ$ is isosceles, because it has two congruent sides, and obtuse, because $\angle X$ is greater than 90°.

Right triangles

In a right triangle, the two sides that form the right angle are called **legs**, and the remaining side, which sits opposite the right angle, is called the **hypotenuse**. In any triangle, the longest side is opposite the largest angle, so the hypotenuse is always the longest side of a right triangle. (See Figure 11-2.)

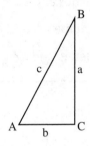

Figure 11-2

The sides of a right triangle always fit the **Pythagorean theorem**, a rule that says that if a and b stand for the lengths of the legs and c stands for the length of the hypotenuse, $a^2 + b^2 = c^2$. For example, if you know the legs of a right triangle measure 8 centimeters and 6 centimeters, the hypotenuse can be found using the Pythagorean theorem.

$$a^2 + b^2 = c^2$$
$$8^2 + 6^2 = c^2$$
$$64 + 36 = c^2$$
$$100 = c^2$$
$$c = \sqrt{100} = 10$$

Equilateral and isosceles triangles

In a scalene triangle, each angle is a different size, but an equilateral triangle is equiangular. It has three congruent sides and three congruent angles, each measuring 60°.

If a triangle is isosceles, it has two congruent sides and two congruent angles, and those angles sit opposite the congruent sides. The two equal sides are called the legs, and the third side is the **base**.

The base angles are the angles formed where each leg meets the base. The base angles of an isosceles triangle are congruent. The remaining angle, formed by the two equal sides, is the vertex angle.

EXERCISE
11·1

In questions 1 through 5, classify each triangle by sides and angles.

1. △ABC with ∠A = 40°, ∠B = 50°, AC = 8 centimeters, and BC = 5 centimeters.

2. △XYZ with ∠Y = 20°, XY = 9 inches, and YZ = 9 inches.

3. △RST with RS = 6 centimeters, ST = 7 centimeters, RT = 8 centimeters. ∠R = 60°, ∠S = 65°, and ∠T = 55°.

4. △DOG with DO = 5 inches, OG = 5 inches, and DG = 5 inches.

5. △CAT with ∠C = 50°, ∠A = 110°, and ∠T = 20°, CA = 4 centimeters, AT = 6 centimeters, and CT = 15 centimeters.

6. △SEA is an isosceles triangle with SE = EA. The measure of ∠S is 35°. Find the measure of ∠E.

7. △BAY is an isosceles triangle with BA = AY. The measure of ∠A is 100°. Find the measure of ∠B.

In questions 8 through 10, use the Pythagorean theorem to find the missing side of the right triangle.

8. △ABC with right angle ∠B has AB = 5 inches and BC = 12 inches. Find AC.

9. △RST with right angle ∠S has RS = 7 centimeters and RT = $7\sqrt{2}$ centimeters. Find ST.

10. △XYZ with right angle ∠Y has XY = 6 meters and XZ = 12 meters. Find YZ.

Quadrilaterals

A quadrilateral is a polygon with four sides. The four angles of a quadrilateral have measurements that total 360°. Within the family of quadrilaterals, different shapes get special names according to their distinct characteristics.

Trapezoids

A trapezoid is a quadrilateral with exactly one pair of parallel sides. The parallel sides are called the bases and the non-parallel sides are legs. The pair of angles formed by one leg and the two bases are **consecutive** angles, and consecutive angles are supplementary. (See Figure 11-3.) Every trapezoid has two pairs of supplementary angles, one pair at the ends of each leg.

If the legs of a trapezoid are congruent, the trapezoid is an isosceles trapezoid. The angles at each end of one of the parallel sides are called base angles. Base angles of an isosceles trapezoid are congruent. The diagonals of an isosceles trapezoid are congruent, but the diagonals of a trapezoid that is not isosceles are not congruent.

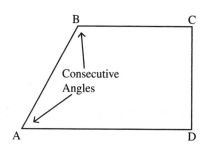

Figure 11-3

The line segment that connects the midpoints of the legs of a trapezoid is called the **median** of the trapezoid. (Sometimes it's called the midsegment.) The median is parallel to the bases, and its length is the average of the lengths of the parallel bases. This is illustrated in Figure 11-4.

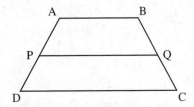

Figure 11-4

In this figure, P is the midpoint of side AD, and Q is the midpoint of side BC in trapezoid $ABCD$. Median PQ is parallel to bases AB and DC. If $AB = 4$ centimeters and $DC = 8$ centimeters, then $PQ = \dfrac{4+8}{2} = 6$ centimeters.

Parallelograms

A parallelogram is a quadrilateral with two pairs of opposite sides that are parallel. Opposite sides of a parallelogram are congruent, and opposite angles are congruent as well. Consecutive angles of a parallelogram are supplementary, and the diagonals of a parallelogram bisect each other. See Figure 11-5.

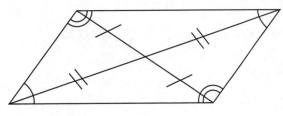

Figure 11-5

Rectangles

A rectangle is a parallelogram that contains a right angle. Since opposite angles are congruent and consecutive angles are supplementary in any parallelogram, the rectangle has four right angles. Every rectangle is a parallelogram and has all the properties of a parallelogram. Its opposite sides are parallel and congruent, its opposite angles are congruent, its consecutive angles are supplementary, and its diagonals bisect each other. In addition, the diagonals of a rectangle are congruent.

Rhombuses

A rhombus is a parallelogram with four congruent sides. A rhombus has all the properties of a parallelogram. In addition, the diagonals of a rhombus are perpendicular, and the diagonals bisect the angles at the vertices they connect.

Squares

A square is a quadrilateral that is both a rectangle and a rhombus. Squares are parallelograms that have four congruent sides and four right angles. They have all the properties of parallelograms, all the properties of rectangles, and all the properties of rhombuses.

ABCD is a trapezoid with $\overline{AB}$ parallel to $\overline{DC}$. Find the measurement of the indicated angle.

1. $\angle A = 84°$ and $\angle C = 63°$. Find $\angle D$.

2. $\overline{AD} \cong \overline{BC}$, $\angle D = 73°$. Find $\angle C$.

3. $\overline{AD} \cong \overline{BC}$, $\angle C = 65°$. Find $\angle A$.

ABCD is a trapezoid with $\overline{AB}$ parallel to $\overline{DC}$ and median $\overline{PQ}$. Find the length of the indicated segment.

4. $AB = 12$ centimeters, $CD = 28$ centimeters. Find PQ.

5. $AB = 8$ inches, $PQ = 15$ inches. Find CD.

Tell whether the quadrilateral WXYZ is a parallelogram based on the information given, or say that it cannot be determined.

6. $\overline{WX}$ is parallel to $\overline{YZ}$ and $\overline{XY}$ is parallel to $\overline{WZ}$.

7. $\angle W = 83°$, $\angle X = 97°$, and $\angle Z = 96°$.

8. $WX = YZ$ and $XY = WZ$

9. $WX = XY$ and $YZ = WZ$

ABCD is a parallelogram. Find the measurement of the indicated angles.

10. $\angle A = 52°$. Find $\angle B$, $\angle C$, and $\angle D$.

11. $\angle A = 93°$. Find $\angle B$, $\angle C$, and $\angle D$.

ABCD is a parallelogram. Find the length of the indicated segment.

12. $AB = 12$ centimeters and $BC = 8$ centimeters. Find CD.

13. $AB = 4$ inches and $BC = 9$ inches. Find AD.

Classify each quadrilateral as parallelogram, rectangle, rhombus, or square.

14.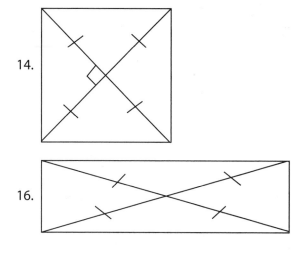

15.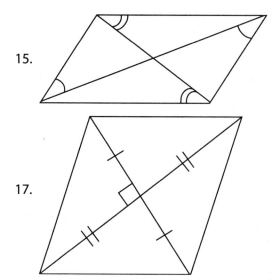

16.

17.

Find the length of the indicated segment.

18. *ABCD* is a square with *AB* = 12 centimeters. Find the length of diagonal *AC*.

19. *PQRS* is a rhombus with diagonals *PR* and *QS* intersecting at *T*. *QR* = 13 inches and *QT* = 5 inches. Find *RT*.

20. *WXYZ* is a rectangle. *WX* = 4 meters and *XY* = 3 meters. Find *XZ*.

21. *JKLM* is a rhombus with diagonals *JL* and *KM* intersecting at *N*. *JL* = 18 centimeters and *KL* = 15 centimeters. Find *MK*.

Find the measurement of the indicated angle.

22. *PQRS* is a rectangle with diagonal *PR*. ∠*PRS* = 28°. Find ∠*SPR*.

23. *ABCD* is a rhombus with diagonals *AC* and *DB* intersecting at *E*. Find ∠*AED*.

24. *JKLM* is a square with diagonals *JL* and *KM* intersecting at *N*. Find ∠*NJK*.

25. *WXYZ* is a rhombus with diagonal *XZ*. ∠*XZY* = 53°. Find ∠*ZWX*.

Other polygons

Triangles and quadrilaterals are the polygons you'll encounter most commonly, but there are many others. There are several differences between the various polygons, but a few characteristics they share.

Interior angles

The total number of degrees in all the angles of a polygon varies with the number of sides. The three angles of a triangle add to 180°. A polygon with more than three sides can be divided into triangles by drawing all possible diagonals from one vertex. The measures of the angles in all those triangles add up to the measures of the interior angles of the polygon. A quadrilateral divides into two triangles; its interior angles add to 2 times 180°, or 360°. In general, the number of triangles created is two less than the number of sides, so the sum of the interior angles of a polygon with n sides is $(n - 2) \cdot 180°$.

If the polygon is regular and all the interior angles are equal size, you can divide the total by the number of angles to find the measure of one interior angle. In a regular polygon with n sides, the measure of one interior angle is $\dfrac{(n-2) \cdot 180°}{n}$.

Exterior angles

An exterior angle of a polygon is formed by extending one side of the polygon through one vertex and beyond. The exterior angle formed this way is supplementary to the interior angle at that vertex, because they form a linear pair.

In a triangle, an exterior angle and the adjacent interior angle add to 180°, and the three angles of a triangle add to 180°. As a result, you can show that the measure of an exterior angle of a triangle is equal to the sum of the two remote interior angles. Since the exterior angle is equal to the two remote interior angles combined, it's larger than either one of them.

In any polygon, if one exterior angle is created at each vertex and the measures of these exterior angles are added up, the total is always 360°, no matter how many sides the polygon has. If the polygon is regular, all the interior angles are congruent, and therefore all the exterior angles are congruent. In a regular polygon of n sides, each exterior angle measures $\dfrac{360}{n}$ degrees.

Diagonals

A line segment that connects two nonadjacent vertices, that is, a line segment that is not itself a side but has vertices of the polygon as its endpoints, is a diagonal. The number of diagonals in a polygon depends on the number of vertices.

NUMBER OF VERTICES	NUMBER OF DIAGONALS
3	0
4	2
5	5
6	9

There are n vertices from which to start drawing. Whichever vertex you choose, you can't draw to the starting vertex, or to either of the vertices adjacent to your starting point, so there are $n - 3$ vertices to which you can draw diagonals. There are $n - 3$ diagonals from each of the n vertices, so there are a total of $n(n - 3)$ diagonals. You need to cut that in half, however, because the diagonal from A to E and the diagonal from E to A, for example, are actually the same line segment, and shouldn't be counted separately. The number of diagonals you can draw is $\dfrac{n(n-3)}{2}$.

To find the number of diagonals in a decagon, remember that it has 10 sides and 10 vertices. From each vertex, 7 diagonals could be drawn, which would seem to say there are 70 diagonals, but you know you're counting each diagonal twice. To get rid of the duplication, you have to divide 70 by 2. There are 35 diagonals in a decagon.

EXERCISE
11·3

For questions 1 through 5, find the total number of degrees in the interior angles of the polygon described.

1. Pentagon

2. Octagon

3. Hexagon

4. Decagon

5. 18-gon

For questions 6 through 10, find the measure of one interior angle of a regular polygon with n sides.

6. $n = 8$

7. $n = 12$

8. $n = 6$

9. $n = 20$

10. $n = 5$

For questions 11 through 15, find the specified measurement.

11. An exterior angle of a regular hexagon

12. The exterior angle at the fifth vertex of a pentagon, if the exterior angles at the other four vertices measure 70°, 74°, 82°, and 61°

13. An exterior angle of a regular polygon with 15 sides

14. The number of diagonals in an octagon

15. The number of diagonals in a polygon with nine sides

Perimeter, circumference, and area

The most fundamental ideas in geometry—point, line, and plane—combine into straight-sided figures known as polygons. These plane figures are flat, two-dimensional shapes, and often you'll want to find their perimeters or their areas. The **perimeter** of a polygon or the **circumference** of a circle measures the linear distance around the figure, and the area measures the space enclosed by a figure. Perimeter can be found by simply adding up the lengths of the sides, whether the polygon has 3 sides or 30. Because perimeter is the sum of the sides, it is measured in the same units used to measure the sides.

The perimeter of a triangle is the sum of the lengths of its sides. If those sides are labeled a, b, and c, you can write a formula $P = a + b + c$. The perimeter of a rectangle is the sum of the lengths of the sides. Since the opposite sides are congruent, the perimeter is twice the length plus twice the width or $P = 2l + 2w$. The perimeter of a square is four times any side, or $P = 4s$, because all sides are congruent. In general, to find a perimeter, just add up the sides.

Area is a measure of the space a figure encloses. It depends on two dimensions, so it is measured in square units. For most figures, you need to know the measurement of one side, called the base, and you need to know the height, or altitude, of the polygon. The height is always measured perpendicular to the base, so if the polygon has two sides that meet at right angles, as happens in a rectangle or a right triangle, you may be able to use one as the base and one as the height. In many cases, the altitude is a distinct measurement. The area of common figures like triangles and parallelograms can be found with the help of simple formulas.

Parallelograms

The area of a parallelogram is the product of the lengths of a base and the height drawn perpendicular to that base, or $A = bh$. In a parallelogram, any side may be the base. The height is the perpendicular distance from the base to its parallel partner. Do not confuse the height with the length of the adjacent side. In rectangles and squares you can use two sides as base and height because the sides meet at a right angle, but that's not true for parallelograms.

To find the area of parallelogram $ABCD$ shown in Figure 12-1, you need to know the length of side AB and the length of altitude CE, which is perpendicular to side AB. You can find the length of AB by

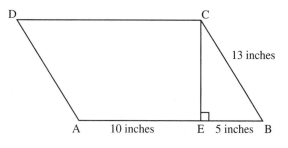

Figure 12-1

adding the length of AE and the length of EB. $AB = 10 + 5 = 15$ inches. To find the length of CE, use the Pythagorean theorem on right triangle $\triangle CEB$.

$$a^2 + b^2 = c^2$$
$$a^2 + 5^2 = 13^2$$
$$a^2 + 25 = 169$$
$$a^2 = 144$$
$$a = 12$$

Parallelogram $ABCD$ has a base of 15 inches and a height of 12 inches, so its area is $A = bh = 15 \times 12 = 180$ square inches.

Rectangles and squares

The area of a rectangle is equal to the product of its length and width, or base and height. The formula for the area of a rectangle is $A = lw = bh$. Because a square is a rectangle with congruent sides, the product of the length and width becomes the length of any side squared, or $A = s^2$.

The area of a rectangle 8 centimeters long and 3 centimeters wide is $A = lw = 8 \times 3 = 24$ square centimeters. A square 5 feet on a side has an area of $A = s^2 = 5^2 = 25$ square feet.

Trapezoids

In a trapezoid, the height is the perpendicular distance between the parallel bases. The area of a trapezoid is half the product of the height and the sum of the bases, or $A = \frac{1}{2}h(b_1 + b_2)$. Since the length of the median that connects the midpoints of the nonparallel sides is half the sum of the bases, the area can also be found by multiplying the length of the midsegment times the height.

In trapezoid $PQRS$ shown in Figure 12-2, side PQ measures 25 inches, side SR measures 17 inches, and altitude XY is 12 inches. Use the formula to find the area.

$$A = \frac{1}{2}h(b_1 + b_2)$$
$$= \frac{1}{2} \times 12(25 + 17)$$
$$= 6 \times 42$$
$$= 252$$

The area of the trapezoid $PQRS$ is 252 square inches.

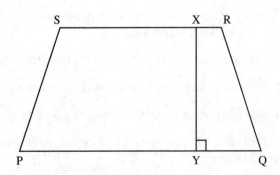

Figure 12-2

Triangles

In a triangle, any side can serve as a base, and the altitude is the perpendicular distance from the opposite vertex to the base. Sometimes, when the triangle has a back-bending shape, caused by an obtuse angle, and an altitude falls outside the triangle, you may have to extend the base to meet the altitude, as shown in Figure 12-3. Don't include the extension in the length of the base.

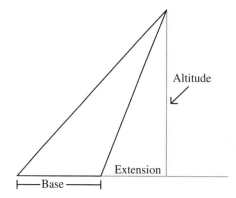

Figure 12-3

The area of a triangle is half the area of a parallelogram with the same base and height. Drawing a diagonal in a parallelogram divides it into two congruent triangles, each with half the area of the parallelogram, so the area of a triangle is half the product of the length of a base and the altitude drawn to that side, or $A = \frac{1}{2}bh$.

If the base of a triangle measures 8 centimeters and the height is 9 centimeters, the area of the triangle is $A = \frac{1}{2}bh = \frac{1}{2} \times 8 \times 9 = 36$ square centimeters.

EXERCISE
12·1

Answer each question based on the information provided.

1. Find the area of a rectangle with a length of 14 centimeters and a width of 8 centimeters.

2. Find the area of a square with a side of 9 meters.

3. Find the width of a rectangle with a length of 13 centimeters and an area of 143 square centimeters.

4. Find the side of a square with an area of 256 square centimeters.

5. The length of a rectangle is 7 centimeters more than its width. If the perimeter of the rectangle is 38 centimeters, what is the area?

6. Find the area of a parallelogram with a base of 18 centimeters and a height of 12 centimeters.

7. Find the area of a trapezoid with bases of 3 meters and 7 meters and a height of 5 meters.

8. Find the area of a rhombus with a side of 14 centimeters and a height of 8 centimeters.

9. Find the area of a trapezoid with a longer base of 35 centimeters and a height and shorter base both equal to 9 centimeters.

10. Find the height of a trapezoid with bases of 31 centimeters and 43 centimeters and an area of 740 square centimeters.

11. Find the area of a triangle with a base of 5 centimeters and a height of 3 centimeters.

12. Find the area of a triangle with a base of 83 centimeters and a height of 42 centimeters.

13. Find the area of an isosceles triangle with a base of 42 centimeters and congruent legs each measuring 35 centimeters. (Hint: Drawing the height perpendicular to the base will create two right triangles, with a hypotenuse of 35 centimeters and a leg of 21 centimeters. You can find the height with the Pythagorean theorem.)

14. Find the area of an equilateral triangle with sides 16 centimeters long.

15. If a triangle with a base of 23 centimeters has an area of 149.5 square centimeters, what is the height?

Regular polygons

Finding the area of a polygon with more than four sides isn't easy to reduce to a formula. Generally, the strategy is to break the polygon into triangles, find the area of each triangle, and add them up. When you do that in a regular polygon, you can work out a general formula.

A polygon is regular if all its sides are congruent and all its angles are congruent. Regular polygons are both equilateral and equiangular. If you drew a circle that passed through all the vertices of a regular polygon, the center of that circle would also be the center of the polygon. The **radius** of a regular polygon is the distance from the center to a vertex. The **apothem** is a segment from the center perpendicular to a side of the polygon. (See Figure 12-4.)

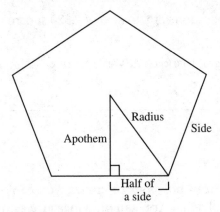

Figure 12-4

The radius, the apothem, and half of a side of the polygon form a right triangle. If you know the measurements of two of the three, you can use the Pythagorean theorem to find the missing one. The angle formed by the radius and the apothem has its vertex at the center of the regular polygon and its measure is $\dfrac{360}{2n} = \dfrac{180}{n}$, where n is the number of sides of the polygon.

To find the area of a regular polygon, you'll need to know the length of a side and the length of the apothem. The area of a regular polygon is equal to half the product of the length of the apothem (a) and the perimeter (P), or $A = \dfrac{1}{2}aP$.

A regular pentagon 8 inches on a side has an apothem 5.5 inches long. The perimeter of the pentagon is $5 \times 8 = 40$ inches. The area of the pentagon, then, is $A = \dfrac{1}{2}aP = \dfrac{1}{2} \times 5.5 \times 40 = 110$ square inches.

Use the formula for the area of a regular polygon to solve each problem.

1. Find the area of a regular hexagon with a perimeter of 84 centimeters and an apothem of $7\sqrt{3}$ centimeters.

2. Find the area of a regular octagon with a side of 10 centimeters and an apothem of 12 centimeters.

3. Find the area of a regular pentagon with a radius of 13 centimeters and a side of 10 centimeters.

4. Find the area of a regular decagon with an apothem of 19 centimeters and a radius of 20 centimeters.

5. Find the area of an equilateral triangle with a radius of 4 centimeters and a side of $4\sqrt{3}$ centimeters.

6. Find the apothem of a regular hexagon with an area of $1,458\sqrt{3}$ square centimeters and a perimeter of $108\sqrt{3}$ centimeters.

7. Find, to the nearest centimeter, the apothem of a regular octagon with an area of 2,028 square centimeters and a side of 21 centimeters.

8. Find the perimeter of a regular pentagon with an area of 292.5 square centimeters and an apothem of 9 centimeters.

9. Find the side of a regular decagon with an area of 2,520 square centimeters and an apothem of 28 centimeters.

10. Find the side of an equilateral triangle with an area of $36\sqrt{3}$ square centimeters and an apothem of $2\sqrt{3}$ centimeters.

Circles

When you shift your focus from polygons to circles, you can't talk about the length of sides or height as you did with the polygons. You can only measure a circle by its radius or diameter, and because the circle is perfectly symmetric, its "width" and its "height" are the same thing.

The circumference, or linear distance around the circle, is the product of π and the diameter, or 2 times the product of π and the radius. As a formula, that becomes $C = \pi d$ or $C = 2\pi r$. In a polygon, the distance around is the perimeter, but in a circle, it's called the circumference. If necessary, the value of π can be approximated as 3.14 or $\dfrac{22}{7}$, but since exact answers are preferred to approximations, you should leave answers in terms of π unless there's a compelling reason to use an approximation.

If a circle has a radius of 35 centimeters, its diameter is 70 centimeters, and its circumference is 70π, or $70 \times \dfrac{22}{7} = 220$ centimeters.

The area of a circle is the product of the constant π and the square of the radius, or $A = \pi r^2$.

Here again, leave answers in terms of π, or approximate using 3.14 or $\dfrac{22}{7}$.

To find the area of the circle mentioned above, with the radius of 35 centimeters, we apply the formula:

$$A = \pi r^2$$
$$= \pi \times 35^2$$
$$= 1{,}225\pi$$
$$\approx 3{,}850$$

The circle has an area of approximately 3,850 square centimeters.

A sector is a portion of a circle cut off by a central angle, like a slice of cake or pie or pizza. An arc is a portion of the circumference of a circle, again, cut off by a central angle. The **measure** of the central angle and the **measure** of the arc it intercepts are the same, and both are measured in degrees. The **length** of the arc depends on the size of the circle as well as the angle. The length of the arc is a fraction of the circumference of the circle, and the fraction is the measure of the central angle over 360, the degrees in a full rotation. So the length of an arc is $S = \dfrac{\text{angle measure}}{360} \cdot \pi d$.

If a central angle of 18° is drawn in a circle with a diameter of 20 centimeters, the length of the arc of that central angle is $S = \dfrac{18}{360} \cdot \pi \cdot 20 = \pi \approx 3.14$ centimeters.

The area of a sector of a circle is a fraction of the area of the entire circle. The fraction is again determined by the central angle of the sector, so the area of a sector is $A = \dfrac{\text{angle measure}}{360} \cdot \pi r^2$.

If a pizza with a radius of 9 inches is cut into 8 equal slices, each slice is a sector with a central angle of $\dfrac{360°}{8} = 45°$. The area of the entire pizza is 81π square inches, so the area of the 45° sector is $A = \dfrac{45}{360} \cdot \pi \cdot 9^2 = \dfrac{1}{8} \cdot 81\pi \approx 31.8$ square inches.

EXERCISE
12·3

Use the formulas for area, circumference, area of a sector, or length of an arc to solve each problem.

1. Find the area of a circle with a radius of 5 centimeters.

2. Find the area of a circle with a diameter of 22 meters.

3. Find the area of a circle with a circumference of 18π centimeters.

4. Find the radius of a circle with an area of 49π square meters.

5. Find the diameter of a circle with an area of 121π square centimeters.

6. Find the circumference of a circle with an area of 16π square inches.

7. Find the area of a sector with a central angle of 40° and a radius of 18 centimeters.

8. Find the radius of a circle if a sector with a central angle of 120° has an area of 27π square centimeters.

9. If a circle has a diameter of 28 centimeters, what is the length of the arc intercepted by a central angle of 18°?

10. A central angle of 150° intercepts an arc 20π inches long. Find the radius of the circle.

Transformations

When you began to study arithmetic, you first learned about numbers and then you learned to operate on them, to add, subtract, multiply, and divide. You learned about points in the coordinate plane, and then formed them into lines and other patterns, and found equations to describe those patterns. And you learned about geometric figures: segments, triangles, rectangles, and other polygons. You learned how to gather information like perimeter and area, and you learned about ways to compare two polygons, looking for similarity or congruence. But how do you operate on those polygons? When does something happen to them?

One way to have an effect on geometric figures is to apply transformations. Transformations are rules that move or change the figure in a particular fashion. There are four common transformations: reflection, translation, rotation, and dilation. Combining these basic elements can create new transformations but these give the essential changes of flip, slide, turn, and resize.

Reflections

Reflection is the most basic transformation. It flips a figure over a line, called the reflecting line. That can actually make quite a few different changes, by changing the direction of the reflecting line or its position in relation to the figure you're reflecting. The figure below shows reflection of a triangle over a vertical line, with very different results depending on where the line is placed.

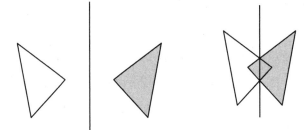

Measure and draw

A reflection moves every point of a geometric figure to a point on the other side of the reflecting line. The original figure is called the preimage, and the figure after reflection is called the image. To find the image of a point:

- ◆ Draw a line from the preimage point perpendicular to the reflecting line

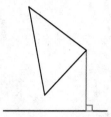

- ◆ Measure the distance along the perpendicular from the preimage point to the reflecting line

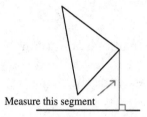

Measure this segment

- ◆ Mark off the same distance along the perpendicular on the other side of the reflecting line

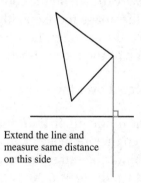

Extend the line and
measure same distance
on this side

- ◆ Mark the image point

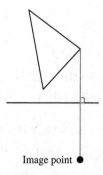

Image point ●

Your protractor can make this a little easier.

Place your protractor so that the center marker and the 90° mark are both on the reflecting line and the preimage point is on the ruler edge of the protractor. Then use the ruler edge to place the image point an equal distance on the other side of the reflecting line.

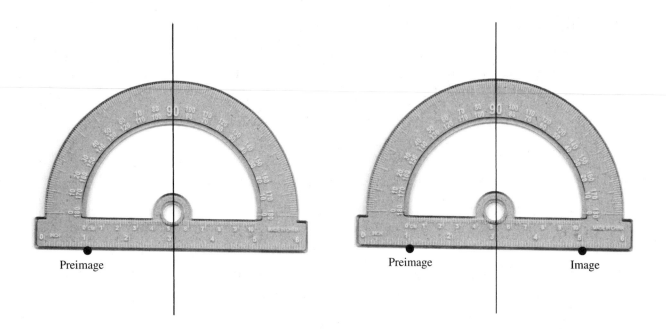

Preimage

Preimage   Image

To reflect a figure with more than one point, simply repeat this process for each vertex of a polygon or each endpoint of a segment. It's not necessary (nor possible) to reflect every point of the figure, just a few key points that define the shape.

Find coordinates and graph

If your preimage and reflecting line are drawn on the coordinate plane, you may be able to use the coordinates of the vertices to find the image more quickly.

If the reflecting line is a vertical line, only the x-coordinates will change. If you reflect a point over the y-axis ($x = 0$), the sign of the x-coordinate changes.

If $(-3, 7)$ is reflected over the y-axis, its image is $(3, 7)$.

If the reflecting line is a vertical line other than the y-axis, a line with an equation of the form $x = c$, find the distance between the preimage (a, b) and the reflecting line $x = c$ by subtracting $c - b = d$, and place the image point the same distance on the other side of the reflecting line. The coordinates of the image point will be $(a + d, b)$.

To reflect $(5, 4)$ over the line $x = -2$,

- Subtract $-2 - 5 = -7$
- Add $-2 + -7 = -9$
- Image point is $(-9, 4)$

If the preimage point is on the reflecting line, it doesn't move.
The image is the same as the preimage.

If the reflecting line is a horizontal line, only the y-coordinates will change, but the changes are similar to those for reflecting over a vertical line. If a point is reflected over the x-axis ($y = 0$), the sign of the y-coordinate changes.

If the point $(6, -1)$ is reflected over the x-axis, its image is $(6, 1)$.

If a point (a, b) is reflected over the horizontal line $y = c$, subtract $c - b = d$ and add d to the y-coordinate. The image point is $(a, b + d)$.

To reflect (−8, 7) over the line $y = 4$,

- Subtract $4 − 7 = −3$
- Add $4 + −3 = 1$
- Image point is (−8, 1)

Reflecting over slanted or oblique lines introduces many more variations, but there are two possible reflecting lines for which the results are easy to anticipate. If the reflecting line is $y = x$, the x-coordinate and the y-coordinate of the preimage will change places to make the image point. The preimage (a, b) becomes the image (b, a). If the reflecting line is $y = −x$, the coordinates of the preimage will change places and change signs.

If the point (−5, 9) is reflected over $y = x$, its image is (9, −5).
If (6, −2) is reflected over the line $y = −x$, the image is (2, −6).

Reflection is called a rigid transformation because the preimage does not bend or stretch or change in any way except to flip over the reflecting line. The image is congruent to the preimage, same shape and same size. The lengths of line segments do not change and the sizes of angles do not change when a figure is reflected.

EXERCISE

13·1

In exercises 1 through 5, the reflecting line and the preimage are shown. Find the image under a reflection.

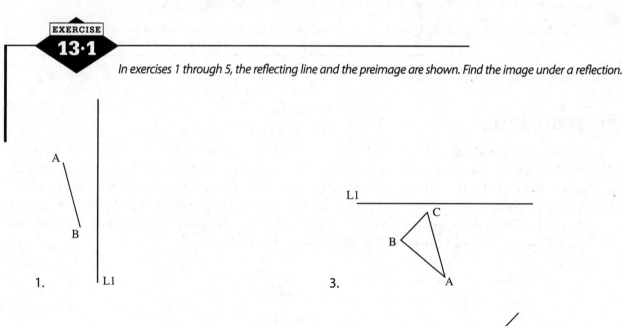

1.

2. L1

3.

4. L1

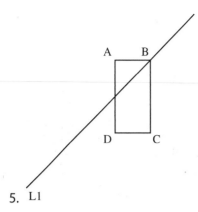

5. L1

In exercises 6 through 10, find the vertices of the image under the reflection described.

6. Line segment $\overline{AB}$ with A(−1, 2) and B(−3, −2) over the line x = 0, the y-axis.

7. Triangle ΔRST with R(2, 2), S(6, −1), and T(1, −3) over the line y = 0, the x-axis.

8. Rectangle □ABCD with A(3, 4), B(5, 4), C(5, −1), and D(3, −1) over the line y = x.

9. Triangle ΔXYZ with X(−7, −2), Y(1, 2), and Z(−1, −5) over the line y = −2.

10. Square □PQRS with P(−4, 5), Q(−1, 5), R(−1, 2), and S(−4, 2) over the line x = 3.

Translations

A translation is a shifting or sliding of the figure to a new location. When you look at the preimage and the image, it may look as though the figure slid along a diagonal line, but the movement will be described by a horizontal move and a vertical move. (Of course, it is possible to have just a horizontal translation or just a vertical shift.) The horizontal change will add or subtract a number from the x-coordinate, and the vertical change will add or subtract from the y-coordinate. The horizontal and vertical change may be the same or different.

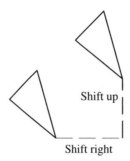

Shift up

Shift right

A translation is actually the result of two reflections, across two different reflecting lines that are parallel to one another. The distance between the reflecting lines determines the distance the figure appears to travel, and the order in which the lines are used controls the direction of the movement: left or right for a horizontal translation, up or down for a vertical shift.

The figure below shows the result of reflecting first over the near line, then over the far line. The final image $\overline{A''B''}$ is to the right of the preimage $\overline{AB}$.

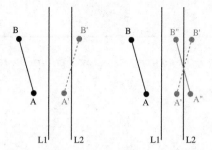

With the same preimage and the same two reflecting lines, you can produce a very different image by reflecting first over the far line then over the line nearer to the preimage. This time, the final image is to the left of the preimage.

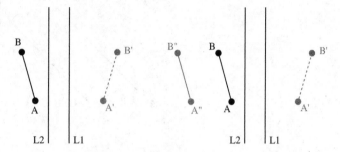

To translate a figure, repeat the translation for each vertex of the figure, as you did with reflection.

Measure and draw

There are two ways you can draw a translation. One is to do the two reflections, using the protractor method you learned in the previous section. The other is simply to measure from the preimage point, left or right the specified distance, and place a point. Then measure from that point, up or down the specified amount to find the final image.

Find coordinates and graph

When you investigate translations on the coordinate plane, you simply need to do arithmetic with the coordinate of the preimage and the size of the desired translation. Because horizontal translations will only change the x-coordinates and vertical translation will only change the y-coordinates, it's easy to combine the two shifts into one new set of coordinates.

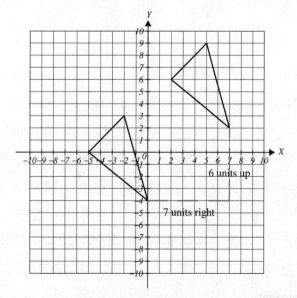

In exercises 1 through 5, the preimage is shown. Perform the reflections, first over line 1 and then over line 2.

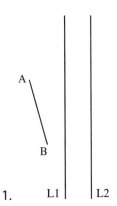

1.   L1   L2

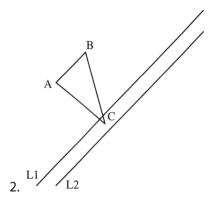

2.

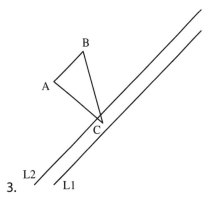

3.

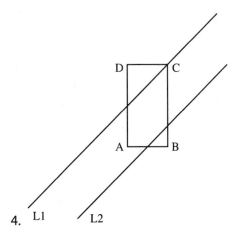

4.   L1   L2

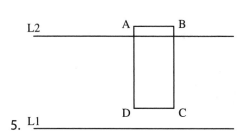

5.

In exercises 6 through 10, sketch the preimage and the image under the translation described.

6. Line segment $\overline{AB}$ with A(−1, 2) and B(−3, −2) shifted 3 units right and 2 units up.

7. Triangle ΔRST with R(2, 2), S(6, −1), and T(1, −3) translated 5 units left and 1 unit down.

8. Rectangle ▭ABCD with A(3, 4), B(5, 4), C(5, −1), and D(3, −1) shifted 3 units left and 4 units up.

9. Triangle ΔXYZ with X(−7, −2), Y(1, 2), and Z(−1, −5) translated 2 units right and 5 units down.

10. Square ▭PQRS with P(−4, 5), Q(−1, 5), R(−1, 2), and S(−4, 2) shifted 3 units left and 3 units up.

Rotations

Imagine that you put a dot on the edge of a wheel, and then spin the wheel around its center. The dot would move, either clockwise or counterclockwise depending on how you spin the wheel. This is a rotation of a point around the center of the wheel.

A rotation is also the result of two reflections over two different reflecting lines, but the lines intersect rather than being parallel. The two reflecting lines form an X and the point where they cross is the center of the rotation, like the center of the wheel. The angle at which the lines cross determines the number of degrees the point rotates, and the order in which the reflecting lines are used determines whether the rotation is clockwise or counterclockwise.

The figure shows reflections over two lines that meet at a 90° angle so the result is that the line segment is rotated 180° counterclockwise.

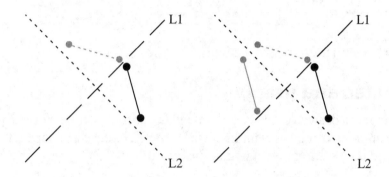

To rotate a figure, repeat the rotation for each of the vertices of the figure.

Measure and draw

Like the translation, a rotation can be drawn in two ways. You can reflect the preimage over one of the lines, and then reflect the result over the other line, using the reflecting technique you learned in Section 13-1.

The order in which the two reflecting lines are used will determine whether the rotation is clockwise or counterclockwise. The measure of the smallest angle formed by the two reflecting lines is half the measure of the final rotation.

If you know whether the rotation is clockwise or counterclockwise, you can measure the amount of rotation and use your protractor to find the image of each point. To find the number of degrees the figure will be rotated, measure the smallest of the vertical angles formed by the reflecting lines. Double that measurement and you have the number of degrees each point of the preimage will move in the rotation.

Place your protractor with the center marker over the center of the rotation, the intersection of the two lines, and the point you want to rotate on the line that connects the center to the 0° mark. Make sure that you have your protractor positioned correctly to rotate clockwise or counterclockwise.

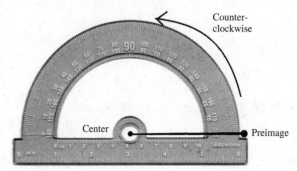

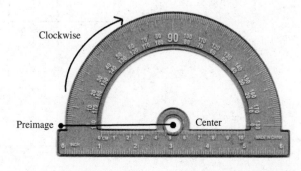

Find the number of degrees in the rotation on the protractor. Mark it with a ray from the center. Measure the distance from the center to the preimage point at 0°, measure the same length on the new ray, and mark the image point.

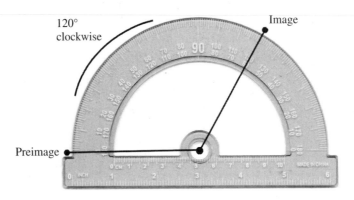

Repeat for each vertex to rotate the figure.

Find coordinates and graph

When the preimage is drawn on the coordinate plane and the center of the rotation is at the origin, common rotations can be accomplished by simple changes to coordinates. If the preimage point is (a, b) and the rotation is centered at the origin:

Preimage	Rotation	Direction	Image
(a, b)	90°	Counterclockwise	$(-b, a)$
(a, b)	90°	Clockwise	$(b, -a)$
(a, b)	180°	Counterclockwise	$(-a, -b)$
(a, b)	180°	Clockwise	$(-a, -b)$
(a, b)	270°	Counterclockwise	$(b, -a)$
(a, b)	270°	Clockwise	$(-b, a)$

If the point $(2, -3)$ is rotated 90° counterclockwise about the origin, its image is $(3, 2)$.

If $(2, -3)$ is rotated 180° counterclockwise, its image is $(-2, 3)$.
If $(2, -3)$ is rotated 270° counterclockwise, its image is $(-3, -2)$.

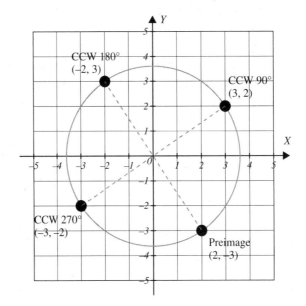

If the point (−4, 2) is rotated 90° clockwise about the origin, its image is (2, 4).

If (−4, 2) is rotated 180° clockwise, its image is (4, −2).
If (−4, 2) is rotated 270° clockwise, its image is (−2, −4).

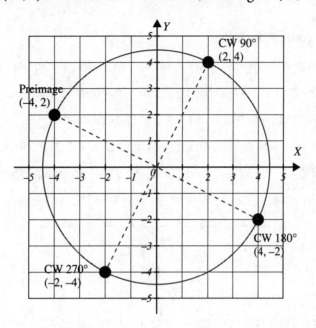

In exercises 1 and 2, the preimage and the two reflecting lines are shown. Perform the rotation by reflecting over line 1 and then over line 2.

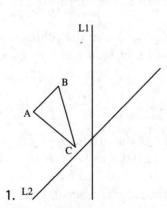

1.

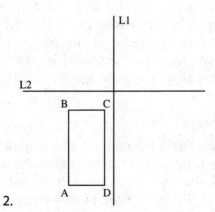

2.

In exercises 3 through 5, the preimage and the center of rotation are shown. Draw the image under the rotation described.

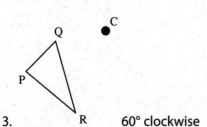

3. 60° clockwise

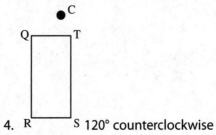

4. 120° counterclockwise

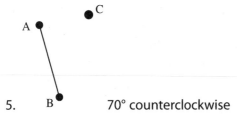

5.   B   70° counterclockwise

In exercises 6 through 10, find the coordinates of the image under the given rotation.

6. Line segment $\overline{AB}$ with A(−1, 2) and B(−3, −2), 90° counterclockwise.

7. Triangle ΔRST with R(2, 2), S(6, −1), and T(1, −3), 180° clockwise.

8. Rectangle ▭ABCD with A(3, 4), B(5, 4), C(5, −1), and D(3, −1), 270° clockwise.

9. Triangle ΔXYZ with X(−7, −2), Y(1, 2), and Z(−1, −5), 270° counterclockwise.

10. Square ▭PORS with P(−4, 5), Q(−1, 5), R(−1, 2), and S(−4, 2), 90° clockwise.

Dilations

Reflections are rigid transformations, and because translations and rotations are created by two reflections, translations and rotations are also rigid transformations. All three of these types of transformations, reflection, translation, and rotation, create images that are congruent to the preimage. Lengths of line segments and measurements of angles are preserved.

Dilation is the one common transformation that is not rigid. It will create an image that is the same shape as the preimage, but the job of a dilation is to change the size. After a dilation, the image may be larger than the preimage, or the image may be smaller. The image will not be congruent to the preimage but it will be similar; same shape, different sizes. The lengths of segments are changed, but angle measurement is not.

There are two pieces of information you will need to perform a dilation: the center, a point outside, inside, or on the preimage, and the scale factor, a number that tells you how much to enlarge or reduce the preimage to create the image.

Enlarging

If the scale factor is a number greater than 1, the dilation will enlarge the preimage to create the image. A scale factor of 2, for example, means that the image will be twice as large as the preimage. Each line segment length will be doubled.

From the center, draw to each vertex of the preimage and extend the line beyond the vertex. If the center is a point of the preimage, the lines you draw may sit right on top of the sides of the preimage, and that's fine.

Measure the distance from the center to a vertex. Multiply that distance by the scale factor, in this example, by 2. The image point will be a point on the same line but at the new distance from the center, in this case, twice as far from the center as the preimage point. Repeat for each vertex.

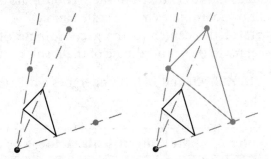

A scale factor of 1 would not change the size at all.
The preimage and the image would be exactly the same size.

Shrinking

When the scale factor is less than 1, the dilation will produce an image smaller than the preimage. The process of drawing the image is the same, but the image, rather than being farther from the center than the preimage, will be closer. When you draw the lines from the center to each vertex, it will not be necessary to extend them beyond the preimage. Because the scale factor is less than 1, when you multiply the distance by the scale factor, the result will be a smaller number, a shorter distance, and the image point will fall on the line somewhere between the center and the preimage.

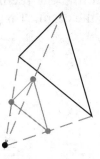

Find coordinates and graph

If the preimage is drawn on the coordinate plane, and the dilation is centered at the origin, finding the coordinates of the image is as simple as multiplying the coordinates of each vertex by the scale factor. If the line segment AB with A(2, −3) and B(3, 1) is the preimage in a dilation with a scale factor of 3, the vertices of the image are A′(6, −9) and B′(9, 3).

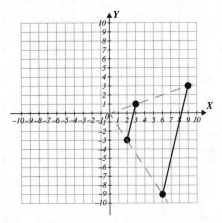

If the dilation is centered at some point other than the origin, follow these steps:

- ◆ Subtract the *x*-coordinate of the center from the *x*-coordinate of the preimage point and subtract the *y*-coordinate of the center from the *y*-coordinate of the preimage.
- ◆ Multiplying the resulting coordinates by the scale factor.
- ◆ Add the *x*-coordinate of the center to the *x*-coordinate of the result and add the *y*-coordinate of the center to the *y*-coordinate of the result.

Suppose line segment $\overline{AB}$ with A(2, −3) and B(3, 1) undergoes a dilation with a scale factor of $\frac{1}{2}$ centered at (−1, 2).

Preimage	Subtract center	Multiply by scale factor	Add center
A(2, −3)	$(2 + 1, -3 - 2) = (3, -5)$	$\left(3 \cdot \frac{1}{2}, -5\frac{1}{2}\right) = \left(\frac{3}{2}, -\frac{5}{2}\right)$	$\left(\frac{3}{2} - 1, -\frac{5}{2} + 2\right) = \left(\frac{1}{2}, -\frac{1}{2}\right)$
B(3, 1)	$(3 + 1, 1 - 2) = (4, -1)$	$\left(4 \cdot \frac{1}{2}, -1\frac{1}{2}\right) = \left(2, -\frac{1}{2}\right)$	$\left(2 - 1, -\frac{1}{2} + 2\right) = \left(1, \frac{3}{2}\right)$

The image of $\overline{AB}$ under a dilation with a scale factor of $\frac{1}{2}$ centered at (−1, 2) is the line segment $\overline{A'B'}$ with A$'\left(0, \frac{1}{3}\right)$ and B$'\left(\frac{1}{3}, \frac{5}{3}\right)$.

The first graph shows the center and the preimage. In the second graph, the problem is translated so that the center moves to the origin. That is accomplished by subtracting the coordinates of the center from each point.

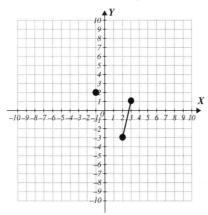
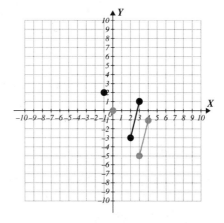

The dilation is performed on the shifted version and then shifted back to the original position.

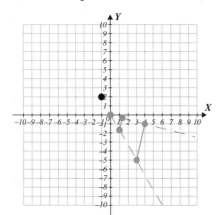
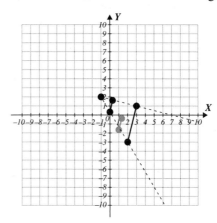

In exercises 1 through 5, the preimage and center C are shown and the scale factor is given. Draw the image.

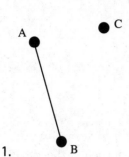

1. Scale factor = $\frac{1}{2}$

4. Scale factor = 3

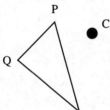

2. Scale factor = 2

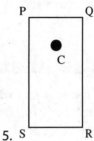

5. Scale factor = 2.5

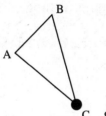

3. Scale factor = 1.5

In exercises 6 through 8, the vertices of the preimage and the scale factor are given. The dilation is centered at the origin. Find the vertices of the image.

6. Line segment $\overline{AB}$ with A(−1, 2) and B(−3, −2). Scale factor 3.

7. Square □PQRS with P(−2, 6), Q(2, 6), R(2, 2), and S(−2, 2). Scale factor 1.5.

8. Rectangle □ABCD with A(3, 4), B(5, 4), C(5, −1), and D(3, −1). Scale factor $\frac{1}{2}$.

In exercises 9 and 10, the vertices of the preimage are given. The center and scale factor of each dilation are indicated. Find the vertices of the image.

9. Line segment $\overline{AB}$ with A(1, 2) and B(−3, −2). Center: (4, 1), scale factor 2.

10. Triangle ΔRST with R(2, 2), S(6, −1), and T(1, −3). Center: (0, 4), scale factor $\frac{1}{2}$.

Circles

A **circle** is the set of all points at a fixed distance from a given point. The given point is called the **center** of the circle, and the fixed distance is the **radius**. The word *radius* is also used to denote a line segment that connects the center to a point on the circle.

Lines and segments in circles

A **chord** is a line segment whose endpoints are points of the circle. A **diameter** is a chord that passes through the center of the circle, the longest chord of the circle. A **secant** is a line that contains a chord. If you extend a chord past the edges of the circle to turn it from a line segment into a line that goes on forever, you form a secant. Secants are lines that intersect the circle in two points. A **tangent** is a line that intersects the circle in exactly one point. It just touches the circle, but doesn't cut through. The point at which the tangent touches the circle is called the **point of tangency**.

EXERCISE
14-1

For questions 1 through 5, use the figure to fill in the blank with one of the words in the list.

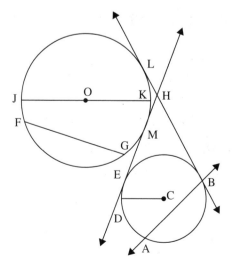

Center Chord Circle Diameter
Radius Secant Tangent

1. $\overleftrightarrow{AB}$ is a _____ of circle C.

2. $\overline{JK}$ is a _____ of circle O.

3. $\overline{FG}$ is a _____ of circle O.

4. $\overline{CD}$ is a _____ of circle C.

5. $\overleftrightarrow{LB}$ is a _____ of circle O.

For questions 6 through 10, identify each of the following elements in the figure.

6. Diameter

7. Tangent

8. Radius

9. Chord

10. Secant

Angles and arcs

When radii, diameters, chords, secants, and tangents intersect one another, they form angles. The measure of an angle formed by such lines and segments depends on which segments intersect, where they intersect, and the size of the arcs, or portions of the circle, that they cut off.

If you imagine the circle divided into 360 tiny sections, called **degrees**, then an **arc** is a portion of the circle and the measure of the arc is the number of degrees it contains. A half circle, or semicircle, contains 180 degrees. The symbol for degree is °, so 180 degrees is written 180°. A quarter of the circle is an arc of 90°.

Vertex at the center: angle = arc

A **central angle** is an angle whose vertex is the center of the circle and whose sides are radii. In Figure 14-1, $\angle NOQ$ is a central angle in circle O. The measure of a central angle is equal to the measure of its intercepted arc. The measure of $\angle NOQ$ is equal to the measure of arc NQ. Arc NQ can be written as $\overset{\frown}{NQ}$.

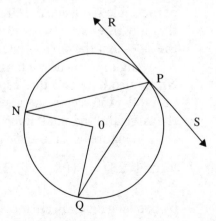

Figure 14-1

Vertex on the circle: angle = $\frac{1}{2}$ arc

An **inscribed angle** is an angle whose vertex is a point on the circle and whose sides are chords. The measure of an inscribed angle is equal to half the measure of its intercepted arc. In Figure 14-1, ∠NPQ is an inscribed angle. It intercepts arc NQ, so the measure of ∠NPQ is half the measure of arc NQ.

An angle formed when a chord meets a tangent at the point of tangency is equal to half the measure of the intercepted arc. If the chord is a diameter, the arc is a semicircle and the angle is a right angle. In Figure 14-1, ∠NPR and ∠QPS are angles formed by a tangent and a chord. The measure of ∠NPR is half of arc NP and the measure of ∠QPS is half of arc QP.

Vertex inside the circle: angle = $\frac{1}{2}$ the sum of the arcs

When two chords intersect within the circle, vertical angles are formed. The vertical angles must be congruent, yet each angle in the pair of vertical angles may intercept a different size arc. The measure of each of the angles in a pair of vertical angles formed when two chords cross is the average of the two intercepted arcs. In Figure 14-2, $\angle 1 = \angle 3 = \frac{1}{2}(\overset{\frown}{AD} + \overset{\frown}{BC})$ and $\angle 2 = \angle 4 = \frac{1}{2}(\overset{\frown}{AC} + \overset{\frown}{BD})$.

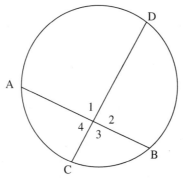

Figure 14-2

Vertex outside the circle: angle = $\frac{1}{2}$ the difference of the arcs

When an angle is formed by two secants drawn from the same point outside the circle, the vertex of the angle is outside the circle and the sides of the angle cut across the circle. Every angle formed this way intercepts two arcs, a small one when the sides first meet the circle and a larger one as the sides exit the circle. The measure of the angle is half the difference of the arcs.

Angles formed by a tangent and a secant, or by two tangents, from a single point outside the circle, also have two intercepted arcs, but because the tangent just touches the circle at a single point, the intercepted arcs will meet each other. In the case of an angle formed by two tangents, the large and the small intercepted arcs together make up the whole circle.

In Figure 14-3, the angle ∠Q is formed by two secants. To find the measure of this angle, use the formula $\angle Q = \frac{1}{2}(\overset{\frown}{PR} - \overset{\frown}{ST})$. The angle formed by a tangent and a secant, ∠Z, and the angle formed by two tangents, ∠B, are measured in similar ways. $\angle Z = \frac{1}{2}(\overset{\frown}{VX} - \overset{\frown}{VY})$ and $\angle B = \frac{1}{2}(\overset{\frown}{ADC} - \overset{\frown}{AC})$.

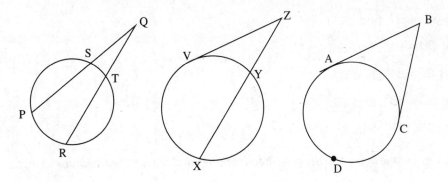

Figure 14-3

Use the figure and the given information to answer questions 1 through 5.

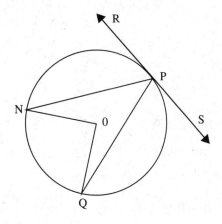

1. If $\overset{\frown}{NQ} = 48°$, find the measure of $\angle O$.

2. If $\overset{\frown}{NQ} = 86°$, find the measure of $\angle NPQ$.

3. If $\angle O = 45°$, find $\overset{\frown}{NQ}$.

4. If $\angle NPQ = 53°$, find $\overset{\frown}{NQ}$.

5. If $\angle O = 28°$, find $\angle NPQ$.

Use the figure and the given information to find the specified measurement in questions 6 through 10.

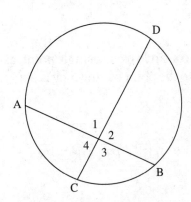

6. If $\overset{\frown}{AD} = 96°$ and $\overset{\frown}{BC} = 62°$, find $\angle 1$.

7. If $\overset{\frown}{AC} = 101°$ and $\overset{\frown}{BD} = 93°$, find $\angle 4$.

8. If $\overset{\frown}{AD} = 118°$ and $\overset{\frown}{BC} = 84°$, find $\angle 3$.

9. If $\overset{\frown}{AC} = 125°$ and $\overset{\frown}{BD} = 63°$, find $\angle 2$.

10. If $\overset{\frown}{AC} = 87°$ and $\overset{\frown}{BD} = 101°$, find $\angle 1$.

In questions 11 through 15, use the figure and the given information to find the specified measurements.

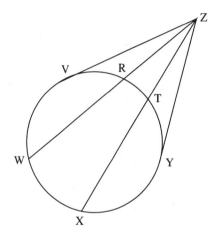

11. If $\overset{\frown}{WX} = 85°$ and $\overset{\frown}{RT} = 23°$, find $\angle WZX$.

12. If $\overset{\frown}{VW} = 93°$ and $\overset{\frown}{VR} = 21°$, find $\angle VZW$.

13. If $\overset{\frown}{VXY} = 285°$, find $\angle VZY$.

14. If $\overset{\frown}{VWX} = 157°$ and $\angle VZX = 57°$, find $\overset{\frown}{VRT}$.

15. If $\overset{\frown}{WX} = 79°$ and $\angle WZX = 33°$, find $\overset{\frown}{RT}$.

Lengths of segments

In addition to forming angles, intersecting line segments in and around the circle also divide one another in predictable ways. Some segments are congruent, while others are proportional, although the rules are usually given as products.

Two tangents

When two tangents are drawn to a circle from a single point, as in Figure 14-4, the tangent segments are congruent. A radius drawn to the point of tangency is perpendicular to the tangent. (The symbol for "is perpendicular to" is ⊥.)

$$PT = PR$$
$$\overline{OT} \perp \overline{PT}, \ \overline{OR} \perp \overline{PR}$$

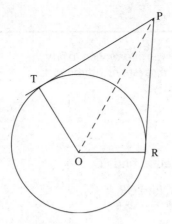

Figure 14-4

Two chords

When two chords intersect in a circle, the chords divide one another in such a way that the product of the lengths of the segments of one chord is equal to the product of the lengths of the segments of the other. In Figure 14-5, $AE \cdot EB = CE \cdot ED$.

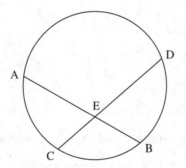

Figure 14-5

If one of the chords is a diameter, and the diameter is perpendicular to the chord, then the diameter bisects the chord and its arc. In Figure 14-6, if you know that diameter $\overline{VW}$ is perpendicular to chord $\overline{XY}$ you know that $\overline{XZ} \cong \overline{ZY}$ and $\overarc{XW} \cong \overarc{WY}$.

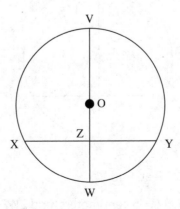

Figure 14-6

Use the figure and the given information to find the specified measurement in questions 1 through 4.

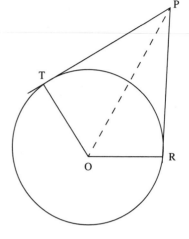

1. If *PT* = 12 centimeters, find *PR*.

2. If *PR* = 28 inches, find *PT*.

3. If *PT* = 12 meters and *OT* = 5 meters, find *PO*.

4. If *PO* = 35 feet and *PR* = 28 feet, find *OR*.

Use the next figure and the given information to find the specified measurement in questions 5 through 7.

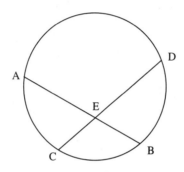

5. If *AE* = 8 centimeters, *EB* = 6 centimeters, and *CE* = 12 centimeters, find *ED*.

6. If *AE* = 9 inches, *ED* = 6 inches, and *CE* = 12 inches, find *EB*.

7. If *CD* = 16 meters, *EB* = 3 meters, and *AE* = 13 meters, find *CE*.

Use the figure and the given information to find the specified measurement in questions 8 through 10.

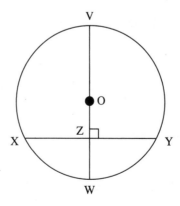

8. If *XZ* = *ZY* = 10 feet and *VZ* = 25 feet, find *VW*.

9. If *XZ* = *ZY* = 3 centimeters and *OV* = 5 centimeters, find *ZW*.

10. If *VZ* = 18 inches and *ZW* = 8 inches, find *XY*.

Volume and surface area

·15·

Three-dimensional objects are commonly called solids, even though they may be hollow. A building block and a cardboard box may have the same shape, but the block is truly solid, while the box has space inside, because it's meant to be filled with something else.

If the solid is made up of shapes like rectangles or triangles or other polygons, it's called a **polyhedron** and each of the polygons is called a **face**. The line segments at which the faces meet are called **edges** and the corners are called **vertices**. (See Figure 15-1.)

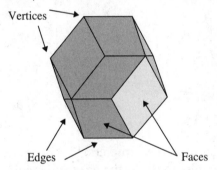

Vertices

Edges   Faces

Figure 15-1

Polyhedra are given names that have to do with the polygons they're built from and how those polygons are put together.

Prisms

A polyhedron with two identical parallel faces, connected by rectangles or parallelograms, is called a **prism**. It is named for the parallel faces. Figure 15-2 shows a hexagonal prism, because the bases—the parallel faces—are hexagons. If the bases were triangles, it would be a triangular prism.

The volume of a polyhedron is a measurement of how much it can hold (if it's hollow) or how much space it takes up. Volume is measured in cubic units. A block that is 1 centimeter wide, 1 centimeter long, and 1 centimeter high has a volume of 1 cubic centimeter.

Solids with straight sides are the easiest to deal with when you're looking for a volume. If you could fill a cardboard box (or other prism) with cubes like that, tightly packed in layers, layer over

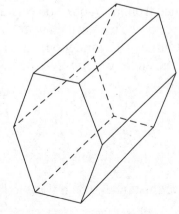

Figure 15-2

layer, then the number of cubes you packed into the box would be its volume. If the box is 20 centimeters long and 15 centimeters wide, you can pack in a layer with 20 rows of 15 one-centimeter cubes, or 300 cubes. If the box is 10 centimeters high, you can make 10 of those layers, so the box will hold 10 × 300, or 3,000 cubes. Its volume is 3,000 cubic centimeters.

The formula for the area of a prism uses B to stand for the area of one of the bases, and h to stand for the height of the prism, the perpendicular distance between the bases. The basic rule for the volume of a prism is to find the area of one of the bases, then multiply by the height of the prism.

$$\text{Volume} = \text{area of the base} \times \text{height}$$
$$V = Bh$$

Imagine a triangular prism that has right triangles as its bases, and suppose it is a right triangular prism, which means that the faces that connect the two bases are perpendicular to the plane of the base (Figure 15-3). In simple terms, if you stand it on its base, it stands up straight, as compared to an oblique prism, which leans to one side.

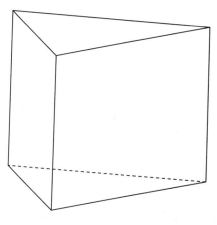

Figure 15-3

The sides of the right triangle measure 5 centimeters, 12 centimeters, and 13 centimeters. You can use the two legs, which form the right angle, as the base and height, so the area of the triangle is $\frac{1}{2}bh = \frac{1}{2} \times 5 \times 12 = 30$ square centimeters. The height of the prism, the distance between the bases, is 8 centimeters. To find the volume, you multiply the area of the base times the height. $V = Bh = 30 \times 8 = 240$ cubic centimeters.

Cylinders

A solid with two parallel bases that are circles is called a **cylinder**. The cylinder is the familiar shape of soft drink cans and most other canned goods. Although a cylinder is not a polyhedron, because it's not formed from polygons, you calculate its volume in a similar fashion. The volume of a cylinder is the area of the base, which is πr^2, times the height. The volume of a cylinder is $V = \pi r^2 h$.

The volume of a cylinder 6 inches high with a radius of 4 inches can be found by calculating the area of the circle and multiplying by the height.

$$V = \pi r^2 h$$
$$= \pi \cdot 4^2 \cdot 6$$
$$= \pi \cdot 16 \cdot 6$$
$$= 96\pi$$

In many circumstances, you'll want to keep the exact value of 96π cubic inches, but you can use an approximate value of π, like 3.14 or $\frac{22}{7}$. $V = 96\pi \approx 96 \times 3.14 \approx 301.44$ cubic inches.

Find the volume of each prism or cylinder.

1. Find the volume of a rectangular prism 30 centimeters high, with a base 14 centimeters long and 8 centimeters wide.

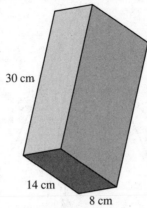

30 cm

14 cm

8 cm

2. Find the volume of a triangular prism 4 inches high, with a base that is a right triangle with legs 3 inches and 4 inches long.

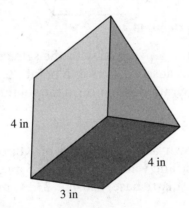

4 in

4 in

3 in

3. The triangle that forms the base of a prism has a base of 8 inches and a height of 12 inches. The prism is 10 inches high. Find its volume.

4. A rectangular prism measures 8 inches long, 5 inches wide, and 9 inches high. Find the volume of the prism.

5. A prism has bases that are hexagons, each of which has an area of 32 square centimeters. If the prism is 15 centimeters high, what is its volume?

6. The octagons that are the bases of a prism each have an area of 85 square inches. The height of the prism is 12 inches. Find the volume of the prism.

7. Find the volume of a cylinder 19 centimeters high with a radius of 8 centimeters.

8. Find the volume of a cylinder with a radius of 15 inches and a height of 9 inches.

9. Find the volume of a cylinder with a diameter of 12 centimeters and a height of 9 centimeters.

10. Find the volume of a cylinder 25 feet high if the circumference of the base is approximately 119.32 feet.

Pyramids

Prisms have two parallel bases; they're "flat" on both ends. Solids that have one base and come to a point at the other end are called **pyramids**. Pyramids get their names from the shape of the base—square pyramids, triangular pyramids—and the other faces are triangles that all tip in to meet at a point. A hexagonal pyramid is illustrated in Figure 15-4.

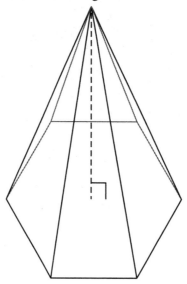

Figure 15-4

If you compare a prism and a pyramid with the same base and height, you'll see that the pyramid takes up less space, because it tapers to a point. Its volume is smaller than that of the prism. In fact, the volume of a pyramid is one-third the volume of a prism with the same base and height, or $V = \frac{1}{3}Bh$.

The square pyramid known as the Pyramid of Khufu, or the Great Pyramid of Giza, measures 230.4 meters on a side and was originally 146.5 meters high. To find the volume of the pyramid, take the area of the square base, 230.4 × 230.4, multiply by the height of 146.5, and take $\frac{1}{3}$ of that. (Remember that multiplying by $\frac{1}{3}$ is the same as dividing by 3.)

$$
\begin{aligned}
V &= \frac{1}{3}Bh \\
&= \frac{1}{3} \times 230.4 \times 230.4 \times 146.5 \\
&= \frac{1}{3} \times 53{,}084.16 \times 146.5 \\
&= \frac{1}{3} \times 7{,}776{,}829.44 \\
&= 2{,}592{,}276.48
\end{aligned}
$$

The volume of the Great Pyramid is 2,592,276.48 cubic meters.

Cones

Just as finding the volume of a cylinder is similar to finding the volume of a prism, finding the volume of a cone is similar to finding the volume of a pyramid. Like the pyramid, the cone has one base—a circle—and comes to a point, and like the pyramid, its volume is one-third the area of the base times the height, or $V = \frac{1}{3}\pi r^2 h$.

To find the volume of a cone with a radius of 7 inches and a height of 12 inches, find the area of the circular base ($\pi \cdot 7^2$), multiply by the height of 12, then multiply by $\frac{1}{3}$ (or divide by 3).

$$V = \frac{1}{3}\pi r^2 h$$

$$= \frac{1}{3}\pi \cdot 7^2 \cdot 12$$

$$= \frac{1}{3}\pi \cdot 49 \cdot 12$$

$$= \frac{1}{3}\pi \cdot 588$$

$$= 196\pi \approx \overset{28}{196} \times \frac{22}{\underset{1}{7}} \approx 616$$

The volume of the cone is approximately 616 cubic inches.

EXERCISE
15·2

Find the volume of each pyramid or cone.

1. Find the volume of a square pyramid 18 inches high if the base measures 15 inches on each side.

2. A triangular pyramid has as its base an equilateral triangle with a side of 8 centimeters. The height of the pyramid is 9 centimeters. Find its volume.

3. A pyramid has a rectangular base 7 feet wide and 9 feet long. If the pyramid is 5 feet high, what is its volume?

4. The base of a hexagonal pyramid has an area of 150 square centimeters. The height of the pyramid is 12 centimeters. Find its volume.

5. A square pyramid 24 feet on each side is 24 feet high. Find its volume.

6. Find the volume of a cone with a radius of 12 inches and a height of 8 inches.

7. Find the volume of a cone with a height of 12 inches and a radius of 8 inches.

8. The circle at the base of a cone has a diameter of 16 centimeters. If the height of the cone is 9 centimeters, what is its volume?

9. The circle at the base of a cone has a circumference of approximately 56.52 meters. If the height of the cone is 15 meters, what is its volume?

10. The square at the base of a pyramid has a perimeter of 24 feet. The pyramid is 8 feet high. The circle at the base of a cone has a circumference of 24 feet. The cone is 8 feet high. Which has the greater volume, the pyramid or the cone?

Spheres

A **sphere** is the three-dimensional equivalent of a circle. It's the shape of many balls, like baseballs and billiard balls (but not footballs). Trying to explain why the formula for the volume of a sphere is what it is can be a tough challenge, but try thinking about orange sections. If you take apart an orange, each section is a wedge, with a shape like a semicircle. Its surface would have the same radius as the orange and would have an area of $\dfrac{\pi r^2}{2}$. The problem is that it is a wedge, not a flat semicircle. What if you had sections that were so thin, so close to flat that you could ignore their thickness, you could use them to approximate the volume of the orange? The number of sections would depend on how thin you could get them, but the thinner you got them, the closer the number would be to the circumference. You'd be approximating the volume by approximately $2\pi r$ wedges each with an area of $\dfrac{\pi r^2}{2}$, which would make your estimate $\pi^2 r^3$. It turns out that's too large, because you can't really ignore the thickness of the wedges, but the volume of a sphere can be found by the formula $V = \dfrac{4}{3}\pi r^3$.

To find the volume of a sphere with a radius of 15 centimeters, calculate

$$V = \frac{4}{3}\pi r^3$$
$$= \frac{4}{3} \cdot \pi \cdot 15^3$$
$$= \frac{4}{3} \cdot \pi \cdot 3{,}375$$
$$= \frac{13{,}500}{3}\pi$$
$$= 4{,}500\pi \approx 14{,}130$$

The volume is $4{,}500\pi$ cubic centimeters, or approximately $14{,}130$ cubic centimeters.

EXERCISE

15·3

For questions 1 through 5, find the volume of the sphere with the given radius.

1. $r = 3$ feet

2. $r = 12$ centimeters

3. $r = 27$ meters

4. $r = 6$ inches

5. $r = 30$ centimeters

For questions 6 through 8, a radius is given. Find the area and circumference of a circle with that radius, and the volume of a sphere with that radius.

6. $r = 15$ feet

7. $r = 48$ centimeters

8. $r = 2$ meters

9. Marjorie's ice-cream shop sells single-scoop ice-cream cones. The cone has a base diameter of 2 inches and a height of 6 inches. Marjorie wants to be certain that if the scoop of ice cream were to melt before it was eaten, the melted ice cream would fit in the cone. Find the volume of the cone. What is the radius of the largest spherical scoop of ice cream that can melt down and still fit in Marjorie's cone?

10. For a class project, Jeffrey is building a scale model of the Unisphere built for the 1964 World's Fair in Flushing Meadows, New York. The scale is 1 inch : 6 feet. The Unisphere is 120 feet in diameter, so Jeffrey plans to build his model with a diameter of 20 inches. Find the volume of the actual Unisphere and the volume of Jeffrey's model. What fraction of the actual volume does the model represent?

Surface area

If you want to wrap a birthday present for a friend, you first have to choose an appropriate box to put the gift in. You need a box with a large enough volume (and a workable shape) to hold the present. When you've boxed the gift and are ready to wrap it, you need to make a rough mental estimate of the surface area of the box. You want to cover the entire box with wrapping paper, so you need to have a sense of the combined area of all the faces of the box.

The surface area of a solid is simple: it is the sum of the areas of all the surfaces. But the surface area can also be complicated, because those surfaces may have many different shapes, even shapes you don't know how to name. As a result, having formulas for surface area can be very helpful.

Prisms and cylinders

To find the surface area of a prism, you need to find the areas of the two parallel bases (or find the area of one base and double it) and then find the areas of the rectangles that are the **lateral faces**. (When someone talks about the lateral area, they're talking about the total of the areas of the lateral faces, without the two bases.)

If you have a rectangular prism, and the dimensions of the bases are l and w, and the height is h, the area of each base is lw, two faces each have an area of lh, and the other two each have an area of wh. The surface area of the rectangular prism is $S = 2lw + 2lh + 2wh$. In the special case of a cube, a rectangular prism in which all faces are squares, the surface area is $S = 6e^2$, where e is the length of an edge.

For a rectangular prism 5 feet long, 3 feet wide, and 4 feet high, use the formula to find the surface area.

$$S = 2lw + 2lh + 2wh$$
$$= 2 \cdot 5 \cdot 3 + 2 \cdot 5 \cdot 4 + 2 \cdot 3 \cdot 4$$
$$= 30 + 40 + 24$$
$$= 94$$

The surface area is 94 square feet.

For other prisms, you will have two bases, which may be triangles or other polygons, and the lateral faces will be rectangles, but the number of lateral faces will depend on the shape of the bases. A triangular prism has three lateral faces, a hexagonal prism has six, and an octagonal prism has eight lateral faces. Rather than creating a different formula for every prism, you can generalize the formula.

The surface area of a prism is equal to twice the area of a base plus the areas of all the lateral faces. $S = 2B + $ total lateral area. The area of each lateral face is equal to the length of an edge of the base times the height of the prism. If you put all those together, the lateral area is equal to the perimeter of the base times the height of the prism. $S = 2B + Ph$, where B is the area of the base, P is the perimeter of the base, and h is the height of the prism.

If a prism has regular hexagons as its bases, and those hexagons have edges 8 inches long, the perimeter of the hexagon is $6 \times 8 = 48$ inches, and the area of the hexagon is $\frac{1}{2} \cdot 4\sqrt{3} \cdot 48 = 96\sqrt{3} \approx 166.28$ square inches. If we know the height of the prism is 10 inches, the surface area can be calculated.

$$S = 2B + Ph$$
$$\approx 2 \cdot 166.28 + 48 \cdot 10$$
$$\approx 332.56 + 480$$
$$\approx 812.56$$

The surface area is 812.56 square inches.

You can apply a similar logic to finding the surface area of a cylinder. The area of one base is πr^2, and doubling that gives you $2\pi r^2$. The lateral area of a cylinder becomes simpler if you think about the cylinder as a can, and the lateral area as the label on the can. When you unroll the label, you have a rectangle. The length of the rectangle is the circumference of the can, and the width of the rectangle is the height of the can. The lateral area is $2\pi r \cdot h$, so the surface area of a cylinder is $S = 2\pi r^2 + 2\pi rh$.

If a cylinder has a radius of 8 centimeters and a height of 9 centimeters, its surface area can be found by applying the formula.

$$S = 2\pi r^2 + 2\pi rh$$
$$= 2\pi \cdot 8^2 + 2\pi \cdot 8 \cdot 9$$
$$= 2\pi \cdot 64 + 2\pi \cdot 72$$
$$= 128\pi + 144\pi$$
$$= 272\pi \approx 854.51$$

The surface area of the cylinder is approximately 854.51 square centimeters.

EXERCISE

15·4

1. Find the surface area of a rectangular prism 18 inches long, 10 inches wide, and 9 inches high.

2. A prism has a base that is a right triangle with legs of 3 inches and 4 inches and a hypotenuse of 5 inches. The height of the prism is 2 inches. Find the surface area of the prism.

3. A prism has bases that are regular pentagons. Each pentagon has an area of 84.3 square centimeters and a perimeter of 35 centimeters. If the height of the prism is 7 centimeters, what is its surface area?

4. A prism with an equilateral triangle as its base measures 4 inches on each base edge, and stands 8 inches high. Find its surface area.

5. Find the surface area of a cube 11 inches on each edge.

6. Find the surface area of a cylinder with a radius of 2 inches and a height of 9 inches.

7. Find the surface area of a cylinder with a diameter of 12 centimeters and a height of 15 centimeters.

8. Find the surface area of a cylinder with a circumference of 314 inches and a height of 14 inches.

9. A rectangular prism has a base 40 inches long and 16 inches wide. The surface area of the prism is 2,512 square inches. What is the height of the prism?

10. A cylinder has a surface area of 357.96 square centimeters and a radius of 6 centimeters. What is the height of the prism?

Pyramids and cones

To talk about the surface area of a pyramid or a cone, you need a measurement you haven't used before. You've measured the base and you've measured the height of the pyramid or cone, but the altitude of the triangles that make up the lateral area is different from the height of the pyramid. The length of the altitude of the lateral faces is called the **slant height**. The letter l is traditionally used to stand for the slant height.

The surface area of a pyramid is the area of the base plus the total of the areas of the lateral faces, each of which is one-half times the edge of the base times the slant height. The lateral area is one-half of the product of the perimeter of the base and the slant height. Therefore, $S = B + \frac{1}{2}Pl$, where B is the area of the base, P is the perimeter of the base, and l is the slant height.

To find the surface area of a triangular pyramid, with an equilateral triangle 6 inches on each side as its base, and a slant height of 4 inches, you'll need to find the area and perimeter of the equilateral triangle (the base). Remember from Exercise 12-1 that the height of an equilateral triangle is half the side times the square root of 3. The triangle that forms the base of the solid has a side of 6 and a height of $3\sqrt{3}$. The area of the base is $\frac{1}{2} \cdot 6 \cdot 3\sqrt{3} = 9\sqrt{3}$ square inches, and the perimeter is 18 inches (6 inches × 3 sides). Apply the formula to find the surface area of the pyramid.

$$S = B + \frac{1}{2}Pl$$
$$= 9\sqrt{3} + \frac{1}{2} \cdot 18 \cdot 4$$
$$= 9\sqrt{3} + 36 \approx 51.59$$

The surface area of the triangular pyramid is approximately 51.59 square inches.

Like the pyramid, the cone has a slant height, the measurement along the sloping side of the cone from the tip to the edge of the base. The surface area of the cone is the area of the base plus the lateral area, or $S = \pi r^2 + \pi r l$, where r is the radius of the cone and l is the slant height.

If a cone has a radius of 2 meters and a slant height of 3 meters, its surface area can be calculated using the formula:

$$S = \pi r^2 + \pi r l$$
$$= \pi \cdot 2^2 + \pi \cdot 2 \cdot 3$$
$$= 4\pi + 6\pi$$
$$= 10\pi \approx 31.4$$

The surface area of the cone is approximately 31.4 square meters.

EXERCISE
15·5

1. Find the surface area of a square pyramid with a base edge of 24 feet and a slant height of 13 feet.

2. Find the surface area of a triangle pyramid if the base is an equilateral triangle 12 centimeters on each side and the slant height is 10 centimeters.

3. The base of a pyramid is a hexagon with an area of approximately 509.22 square inches and a perimeter of 84 inches. The slant height of the pyramid is 16 inches. Find the surface area of the pyramid.

4. The base of an octagonal pyramid has an area of 480 square meters and a perimeter of 192 meters. If the slant height of the pyramid is 30 meters, what is the surface area of the pyramid?

5. Find the surface area of a cone that has a radius of 25 centimeters and a slant height of 32 centimeters.

6. Find the surface area of a cone where the radius is 12 inches and the slant height is 24 inches.

7. Find the surface area of cone with a diameter of 22 centimeters and a slant height of 19 centimeters.

8. A cone has a slant height of 56 inches and the circumference of the base is 201 inches. Find the surface area of the cone.

9. A pyramid has a base that is a regular pentagon with an area of 139.36 square meters. The slant height of the pyramid is 11 meters and the surface area of the pyramid is 1,184.36 square meters. Find the length of a base edge.

10. A cone has a radius of 5 inches and a surface area of 282.75 square inches. Find the slant height of the cone.

Spheres

The surface area of a sphere, like all areas, is measured in square units. If you think about the area of circle of a certain radius, and compare it to the surface of a sphere with the same radius, you can see that the surface area of the sphere is the larger. The circle wouldn't cover even half of the sphere. In fact, the surface area of a sphere is given by the formula $S = 4\pi r^2$. If a sphere has a radius of 5 centimeters, we can apply the formula to find the surface area.

$$S = 4\pi r^2$$
$$= 4\pi \cdot 5^2$$
$$= 4\pi \cdot 25$$
$$= 100\pi \approx 314$$

The sphere has a surface area of approximately 314 square centimeters.

**EXERCISE
15·6**

In questions 1 through 7 the radius of a sphere is given. Find the surface area of the sphere.

1. $r = 2$ inches

2. $r = 5$ centimeters

3. $r = 8$ meters

4. $r = 12$ feet

5. $r = 21$ yards

6. $r = 18$ inches

7. $r = 25$ millimeters

In questions 8 through 10, the surface area of a sphere is given. Find the radius of the sphere.

8. $S = 36\pi$ square meters

9. $S = 64\pi$ square inches

10. $S = 144\pi$ square centimeters

# Counting and probability

Possibly the most important mathematical skills in today's world are those that allow you to make sense of the flood of information and to assess the risks and possibilities in a situation. Making sense of information requires the ability to summarize a lot of information efficiently, present it in clear visuals, and express the relationships. Assessing risk and possibility requires the ability to estimate probabilities, and that in turn requires that you can count the options quickly.

Getting a count

If you camp out all night to get tickets to hear your favorite band in concert, how good are the chances that you'll actually get the tickets? It's pretty much impossible to answer that question until you know how many tickets are actually going on sale and how many people are in line ahead of you. In fact, it's impossible to get a realistic assessment of the probability of any event until you have a count of the number of things that can happen. If there are a lot of possibilities, you'll need strategies for getting an accurate count quickly.

The basic counting principle

When you purchase a new bike lock, you often can set your own lock code. If the lock has three dials and each has 10 digits, you have to choose one of a thousand codes, 000 through 999. There are 10 choices for the first digit, 10 for the second, and 10 for the third. Since you can pair any one of the 10 first digits with any one of the 10 second digits, there are 100 possible pairs. Then any one of those pairs can be completed by any one of the final 10 digits, giving you 100×10, or 1,000 possible codes.

The fundamental rule for quick counting is called the **basic counting principle**. It gives you a quick way of counting up the possibilities by multiplying the number of choices. If you have to do something that requires several choices, and you create a slot for each choice that needs to be made, and fill each slot with the number of options for each choice, multiplying those numbers will tell you how many different possibilities you have.

If you're packing for vacation and don't want to pay fees for extra bags, you'll want to be able to make a lot of different outfits from a few pieces of clothing. If you pack 4 shirts, 3 pairs of slacks, and 2 pairs of shoes, and you're comfortable mixing and matching any of them, how many different outfits can

you make? Using the basic counting principle, create one slot for shirts, one for slacks, and one for shoes: __ __ __. Then fill in the number of each available, and multiply.
_{shirts slacks shoes}

$$\underset{\text{shirts}}{4} \times \underset{\text{slacks}}{3} \times \underset{\text{shoes}}{2} = \underset{\text{outfits}}{24}$$

Now, you may not like all 24 outfits, and you might not be willing to do laundry in your hotel, but you could wear a different outfit every day for more than three weeks with just those few items. Add a couple of sweaters, and you'd be set for more than a month, since

$$\underset{\text{sweaters}}{2} \times \underset{\text{shirts}}{4} \times \underset{\text{slacks}}{3} \times \underset{\text{shoes}}{2} = \underset{\text{outfits}}{48}$$

A popular game gives you a seven-letter word and asks you to find as many smaller words as you can from the letters in the given word. You might, for example, be given the word HISTORY, and then you try to make words from those letters. Some will pop right up, like HIS and STORY, while others, like STY, SOY, HOT, or ROT, require a little rearranging. Some groups of letters won't make a real word. RHST just doesn't spell anything. But how many words are possible? That's a tough question, because there are two-letter words, three-letter words, four-letter words, up to seven-letter words, and because groups of letters like RHST aren't really words. But you could start by counting the number of letter combinations with a certain number of letters and worry later about whether they're sensible words.

How many three-letter "words" can you make from the letters in HISTORY? Well, you have 7 choices for the first letter, but after you use that letter, you can't use it again, so you have 6 choices for the second letter. At that point, you've used two letters, so you have 5 choices for the last letter.

$$\underset{\substack{\text{1st}\\\text{letter}}}{7} \times \underset{\substack{\text{2nd}\\\text{letter}}}{6} \times \underset{\substack{\text{3rd}\\\text{letter}}}{5} = \underset{\text{"words"}}{210}$$

There are 210 different groups of three letters that you can consider to see which ones are really words.

EXERCISE 16·1

1. Claire is making name badges for a conference. She has 2 colors of cards on which to print the badges, 3 fonts in which names can be printed, and 4 colors of ribbon that can be affixed to the badge. How many different badge styles can Claire produce?

2. If Claire counts badges without any ribbon as another style, how many styles can she produce?

3. If you have 8 pairs of slacks, 12 shirts, and 4 sweaters, how many different outfits, each consisting of slacks, a shirt, and a sweater, can you make?

4. A serial number is to be constructed from 2 letters and 5 digits. If both letters and digits can be repeated, how many different serial numbers are possible?

5. How many serial numbers of 2 letters and 5 digits can be formed if letters cannot be repeated but digits can?

6. When asked to create a password, you are often told that it must contain at least 6 characters. How many passwords of 6 different letters are possible with a 26-letter alphabet?

7. How many passwords of 6 different digits are possible with ten digits, 0–9?

8. How many 6-character passwords are possible if you use either letters or numbers?

9. US Social Security numbers are made up of 9 digits, grouped as 3, 2, and 4 (xxx-xx-xxxx). Originally, the first 3 digits indicated the region of the country in which the cardholder lived, but as of 2011, the numbers have been randomized.

   a) How many possibilities are there for the 3-digit group if there were no restrictions on the numbers?
   b) How many possibilities are there for the 2-digit group if there were no restrictions on the numbers?
   c) How many possibilities are there for the 4-digit group if there were no restrictions on the numbers?
   d) How many different 9-digit Social Security numbers would be possible if there were no restrictions on the numbers?

10. The Social Security Administration does have some restrictions on the numbers; specifically, the first 3 digits cannot be 000, or 666, or any 3-digit number beginning with 9 (that is, 900 through 999). How many Social Security numbers are possible with those restrictions? (Hint: This is easier to count if you think about the three parts of the number rather than the individual digits.)

For questions 11 through 15, consider a pizza shop that offers thick or thin crust pizzas that can be topped with meats or vegetables or both. The shop offers sausage, pepperoni, meatballs, or chicken as meat toppings, and peppers, onions, mushrooms, olives, broccoli, tomatoes, eggplant, and zucchini as vegetables.

11. How many thin-crust pizzas with 1 meat and 1 vegetable can you order?

12. How many thick-crust pizzas with no meat and 3 different vegetables can you order?

13. If you like both thin- and thick-crust pizza, and you want 2 kinds of meat and 1 vegetable, how many different pizzas can you order?

14. How many pizzas can be made with 1 meat and 2 different vegetables, if you like both thin and thick crust?

15. The deluxe pizza includes your choice of crust, 3 different meats, and 4 different vegetables. How many ways can you order the deluxe?

Factorials and permutations

If the 15 players on a Little League team line up for hot dogs after the game, in how many different orders can they line up? If it sounds like a basic counting principle question, well, it is. It's a big one—you'll need 15 slots—but you can find the answer with the basic counting principle. You'll probably want a calculator, because

$$\underline{15} \times \underline{14} \times \underline{13} \times \underline{12} \times \underline{11} \times \underline{10} \times \underline{9} \times \underline{8} \times \underline{7} \times \underline{6} \times \underline{5} \times \underline{4} \times \underline{3} \times \underline{2} \times \underline{1} \approx 1.3 \times 10^{12}$$

This pattern of multiplication, from a number—in this case, 15—down to 1, is called a **factorial**. The symbol for a factorial is an exclamation point.

$$15! = \underline{15} \times \underline{14} \times \underline{13} \times \underline{12} \times \underline{11} \times \underline{10} \times \underline{9} \times \underline{8} \times \underline{7} \times \underline{6} \times \underline{5} \times \underline{4} \times \underline{3} \times \underline{2} \times \underline{1} \approx 1.3 \times 10^{12}$$

The number of orders in which the 15 Little Leaguers could line up, the number of different arrangements of 15 people taken 15 at a time, is 15!. An arrangement or ordering is called a **permutation**, and for any number n, the permutations of n things taken n at a time is $n!$.

But what if you're not taking n things at a time? For example, the coach for that Little League team needs to put 9 players on the field. If there are 15 players on the roster, and each of them can play any position, how many different lineups can the coach create?

You're still talking about an arrangement, or permutation, but an arrangement of 9 out of the 15, not all 15. You don't want the entire 15!, but you do want the first 9 slots of it. There are 15 choices for the leadoff hitter, 14 for the second spot, 13 for the third, 12 for the cleanup position, and so on, down to 7 choices for the last spot in the lineup.

$$\underset{\text{1st}}{15} \times \underset{\text{2nd}}{14} \times \underset{\text{3rd}}{13} \times \underset{\text{4th}}{12} \times \underset{\text{5th}}{11} \times \underset{\text{6th}}{10} \times \underset{\text{7th}}{9} \times \underset{\text{8th}}{8} \times \underset{\text{9th}}{7}$$

That product represents the permutations of 15 things taken 9 at a time. It's not the whole 15! It's a truncated version of that factorial. There's no symbol for a truncated factorial, but you can write a formula for the permutations of n things taken t at a time as

$$_nP_t = \frac{n!}{(n-t)!}$$

In this case, the permutations of 15 things taken 9 at a time looks like this:

$$_{15}P_9 = \frac{15!}{(15-9)!}$$
$$= \frac{15!}{6!}$$
$$= \frac{15 \times 14 \times 13 \times 12 \times 11 \times 10 \times 9 \times 8 \times 7 \times \cancel{6 \times 5 \times 4 \times 3 \times 2 \times 1}}{\cancel{6 \times 5 \times 4 \times 3 \times 2 \times 1}}$$
$$= 1{,}816{,}214{,}400$$

Notice that the factorial in the denominator cancels the last six factors of the factorial in the numerator, giving you the truncated version you found with the basic counting principle. So if you need the permutations of 7 things taken 4 at a time, the basic counting principle says you want $\underline{7} \times \underline{6} \times \underline{5} \times \underline{4}$, and the permutation formula says

$$_7P_4 = \frac{7!}{(7-4)!} = \frac{7!}{3!} = \frac{7 \times 6 \times 5 \times 4 \times \cancel{3 \times 2 \times 1}}{\cancel{3 \times 2 \times 1}}$$

EXERCISE

16·2

Evaluate each expression and give your answer in simplest form.

1. $\dfrac{3!}{2!}$

2. $\dfrac{5!}{3!} + \dfrac{6!}{4!}$

3. $8! - 4!$

4. $\dfrac{8! \times 5!}{6!}$

5. $\dfrac{6! \times 4!}{5! \times 7!}$

6. Find the permutations of 8 things taken 5 at a time.

7. Find the permutations of 7 things taken 4 at a time.

8. Find $_9P_8$.

9. Find $_{10}P_2$.

10. Find $_{12}P_1$.

11. How many 4-letter "words" can be formed from the letters in HOCKEY?

12. How many 5-letter "words" can be formed from the letters in FAMILY?

13. How many 3-letter "words" can be formed from the letters in BLANKET?

14. How many 4-letter "words" can be formed from the letters in PYRAMID?

15. How many 6-letter "words" can be formed from the letters in PUNCHBOARD?

Combinations

If 12 people run a race for which medals are given to first-, second-, and third-place finishers, and you're interested in how many different finishing orders are possible, you want the permutations of 12 things taken 3 at a time, or $_{12}P_3 = \dfrac{12!}{(12-3)!} = \dfrac{12!}{9!} = 12 \times 11 \times 10 = 1{,}320$. There are 1,320 different orders of finish.

On the other hand, if you just want to pick a committee of three contestants to design the medals, the order in which you choose them doesn't matter. A committee made up of Alex, Pat, and Sam isn't any different from a committee made up of Pat, Sam, and Alex, or of Sam, Alex, and Pat. It's different from a committee made up of Pat, Gerry, and Lou, or a committee of Alex, Gerry, and Sam, but the same three people should only count once, no matter how you name them.

Because all the different arrangements of the same three people shouldn't count separately, the number of permutations is too big, but it can give you a place to start. There are 1,320 permutations or arrangements of 12 people taken 3 at a time, but fewer combinations or groups to form committees. If you look at a particular group of three—say, Alex, Pat, and Sam—you can probably figure out how to reduce that number. How many different orders of Alex, Pat, and Sam are possible? There should be six. You can calculate that—it's the permutations of 3 things taken 3 at a time—or you can convince yourself with a list.

Alex, Pat, and Sam
Alex, Sam, and Pat
Pat, Alex, and Sam
Pat, Sam, and Alex
Sam, Alex, and Pat
Sam, Pat, and Alex

Since there are six arrangements of these same three people—six arrangements of any three people—the number of permutations is six times larger than the number of groups or combinations. Dividing the number of permutations by 6, or $\dfrac{_{12}P_3}{_3P_3}$, will give you the number of combinations. The combinations of n things taken t at a time is

$$_nC_t = \frac{_nP_t}{_tP_t} = \frac{n!}{(n-t)!} \div t! = \frac{n!}{t!(n-t)!}$$

The permutations of 10 things taken 5 at a time is

$$_{10}P_5 = \frac{10!}{(10-5)!} = \frac{10!}{5!} = \frac{10 \times 9 \times 8 \times 7 \times 6 \times \cancel{5 \times 4 \times 3 \times 2 \times 1}}{\cancel{5 \times 4 \times 3 \times 2 \times 1}} = 30{,}240$$

The combinations of 10 things taken 5 at a time is

$$_{10}C_5 = \frac{_{10}P_5}{_5P_5} = \frac{10!}{5!(10-5)!} = \frac{10\times9\times8\times7\times6\times\cancel{5\times4\times3\times2\times1}}{(5\times4\times3\times2\times1)(\cancel{5\times4\times3\times2\times1})} = \frac{\cancel{10}^2\times9\times\cancel{8}^2\times7\times\cancel{6}}{\cancel{5}\times\cancel{4}\times\cancel{3\times2}\times1} = 252$$

The number of combinations will always be smaller than the number of permutations. Use permutations when the arrangement or order matters, and combinations when a group should only be counted once, no matter how it's arranged.

If you want to choose a three-color palate for your redecorating project, from a collection of 25 paint colors, order doesn't matter, so the number of combinations is

$$_{25}C_3 = \frac{_{25}P_3}{_3P_3} = \frac{25!}{3!(25-2)!} = \frac{25\times\cancel{24}^4\times23\times\cancel{22\times21\times...\times2\times1}}{(\cancel{3\times2}\times1)(\cancel{22\times21\times...\times2\times1})} = 2{,}300$$

There are 2,300 possible color schemes.

EXERCISE 16·3

1. Find the combinations of 6 things taken 3 at a time.

2. Find the combinations of 9 things taken 7 at a time.

3. Find $_{12}C_5$.

4. Find $_5C_5$.

5. Find $_5C_1$.

6. In how many ways can a club with 40 members select 4 officers?

7. How many 5-card poker hands can be dealt from a deck of 52 cards?

8. How many 13-card bridge hands can be dealt from a deck of 52 cards?

9. To buy a certain lottery ticket, a player must choose 5 numbers from the numbers 1 through 56. How many different tickets are possible?

10. How many lottery tickets are possible if the player must choose 6 numbers from the numbers 1 through 56?

Dealing with duplication

Remember that word game that asked you to make words from the letters of a word like HISTORY? With a given word like HISTORY, in which each letter occurs only once, all the permutations are distinct, but if your given word were COOKERY, you'd find that when you tried to list the various arrangements, the duplicate letters—two Os—cause some of the permutations to look exactly the same. Because of the presence of duplicate letters, there are fewer distinct permutations, but again the question is how to cut the number down to size.

Think about the number of 7-letter "words," or the permutation of 7 things taken 7 at a time. The permutations of 7 things taken 7 at a time should be 7!. But CO_1O_2KERY will, without the little numbers to identify the Os, look identical to CO_2O_1KERY. Those two identical letters will cut the number in half. When there is a duplicate item involved in a permutation, the number of distinct permutations of n things taken t at a time can be found with the usual formula, $_nP_t = \dfrac{n!}{(n-t)!}$, divided by the factorial of the number of identical elements. The number of 7-letter "words" that can be formed from the letters in COOKERY is $\dfrac{_7P_7}{2!}$ because there are two identical Os.

$$\frac{_7P_7}{2!} = \frac{7!}{2!} = \frac{7 \times 6 \times 5 \times 4 \times 3 \times \cancel{2 \times 1}}{\cancel{2 \times 1}} = 2{,}520$$

If there is more than one duplicate element, divide by a factorial for each one. The number of 7-letter "words" that can be formed from the letters in BOOKBAG is

$$\frac{_7P_7}{2! \times 2!} = \frac{7!}{2! \times 2!} = \frac{7 \times 6 \times 5 \times 4 \times 3 \times \cancel{2 \times 1}}{2 \times 1 \times \cancel{2 \times 1}} = 1{,}260$$

There's a 2! for the two Bs, and another for the two Os. The number of 10-letter "words" that can be formed from the letters in BOOKKEEPER is

$$\frac{_{10}P_{10}}{2! \times 2! \times 3!} = \frac{10!}{2! \times 2! \times 3!} = \frac{10 \times 9 \times 8 \times 7 \times 6 \times 5 \times 4 \times \cancel{3 \times 2 \times 1}}{2 \times 1 \times 2 \times 1 \times \cancel{3 \times 2 \times 1}} = 151{,}200$$

EXERCISE

16·4

1. Find the number of distinct 8-letter "words" that can be formed from the letters in POLITICS.

2. Find the number of distinct 9-letter "words" that can be formed from the letters in STONEHENGE.

3. Find the number of distinct 7-letter "words" that can be formed from the letters in ABBREVIATION.

4. Find the number of distinct 5-letter "words" that can be formed from the letters in HOUSEWORK.

5. Find the number of distinct 6-letter "words" that can be formed from the letters in INFINITE.

6. Find the number of distinct 4-letter "words" that can be formed from the letters in TENNIS.

7. Find the number of distinct 8-letter "words" that can be formed from the letters in SAILBOAT.

8. Find the number of distinct 10-letter "words" that can be formed from the letters in MISSISSIPPI.

9. Find the number of distinct 5-letter "words" that can be formed from the letters in TELEVISION.

10. Find the number of distinct 4-letter "words" that can be formed from the letters in PERMUTATION.

Probability

Have you ever bought a lottery ticket? Did you win? Did you expect to win? Did you have any idea what your chances of winning were? The ticket probably included information on your chance of winning, but most people who play the lottery don't read that fine print. If they did, they might not be so quick to hand over their money.

Simple probability

The probability of something happening is a number between zero and one that tells you how likely the event is. If the probability is zero, the event is impossible. It can't happen. If the probability is one, it's absolutely certain to happen. It's a sure thing.

Most probabilities are somewhere in between zero and one, because most events are neither absolutely impossible nor absolutely certain. Probabilities are fractions, but you'll often hear them expressed as percents—a 30% chance of rain—or as ratios—1 chance in 10 of such-and-such happening. The probability of an event is a fraction that compares the number of ways the event can happen to the total number of things that can happen. The things that can possibly happen are called outcomes, and if event E happens, that's considered a success.

$$P(E) = \frac{\text{\# of ways E can happen}}{\text{\# of things that can happen}} = \frac{\text{\# of successes}}{\text{\# of outcomes}}$$

If you take a well-shuffled deck of cards and pick one card at random, the probability of choosing an ace is $\frac{4}{52} = \frac{1}{13}$ because there are four aces, so four possible successes, out of 52 cards in the deck. The probability that the card will be a heart is $\frac{13}{52} = \frac{1}{4}$ because 13 of the 52 cards are hearts.

The probability that you will choose the ace of hearts is $\frac{1}{52}$ because there is only one ace of hearts in the deck.

Sometimes you can't count successes and outcomes, but you can measure them. If you're throwing a dart at a dartboard, you can't count the number of places it can land, but you can find the area of the board. You can't count the number of ways the dart can hit the bull's-eye, but you can find the area of the bull's-eye. You can express the probability of hitting the bull's-eye as the area of the bull's-eye over the area of the board. (Of course, hitting a bull's-eye with a dart isn't purely chance. There is skill involved, but this notion of geometric probability does give a way to determine the chance of hitting a particular part of the board.)

EXERCISE
16·5

For questions 1 through 3, one card is drawn at random from a standard deck of 52 cards.

1. What is the probability of drawing a black card?

2. What is the probability of drawing a diamond?

3. What is the probability of drawing a card showing an odd number (3, 5, 7, or 9)?

For questions 4 through 7, a fair die bearing the numbers 1 through 6 is rolled.

4. What is the probability of rolling a 3?

5. What is the probability of rolling an even number?

6. What is the probability of rolling a number less than 7?

7. What is the probability of rolling a 7?

For questions 8 through 10, a single marble is drawn, without looking, from a bag that contains 25 marbles. Five are blue, 6 are red, 4 are green, and 10 are white.

8. What is the probability of drawing a blue marble?

9. What is the probability of drawing a marble that is not white?

10. What is the probability of drawing a yellow marble?

Compound probability: and

Simple probability is pretty simple, but life is complicated. Trying to guess one number that someone is thinking of may be fun for little children, but you're not likely to find a lottery that asks you to pick one number. Most ask you to pick at least three and some as many as six. The probability of guessing the right number between 1 and 100 is simple: $\frac{1}{100}$. The probability of choosing the right six numbers from some group is a bit more complicated.

Any time you need to find the probability of this event *and* that one, you want to multiply the probability of the first event by the probability of the second event. To see why, you need to go back to the basic counting principle and the definition of probability.

Suppose you roll a die, record the number that comes up, then roll again and record the second result. What's the probability that both of them are sixes? Well, start by looking at the number of things that can happen. There are six possibilities for how the die can land on the first throw and six possibilities for the second throw. By the basic counting principle, that's 6 × 6, or 36 possible outcomes. Of those 36, there is only one way to get two sixes, so the probability of getting two sixes is $\frac{1}{36}$. The probability of getting a six on the first roll is $\frac{1}{6}$, and the probability of getting a six on the second roll is also $\frac{1}{6}$, and $\frac{1}{6} \times \frac{1}{6} = \frac{1}{36}$.

If you perform the same experiment, but ask for the probability of getting two even numbers, the total possible outcomes still number 36, but because the die has three even numbers—2, 4, and 6—there are 3 × 3 = 9 ways to get two even numbers. The probability of two even numbers is $\frac{3}{6} \times \frac{3}{6} = \frac{9}{36} = \frac{1}{4}$.

Sometimes, when you look at the probability of two events occurring in sequence, the results of the first event have an effect on the probability of the second. Rolling the die twice is what's called independent events or independent trials. The result of the first roll has no effect on the probabilities for the second. If you choose one card at random from a deck, record what it is, then put it back in the deck and shuffle before you pull a second card, the probabilities for the second draw are unchanged. Drawing two cards with replacement are independent events.

But what if you didn't put that first card back? When you drew the second card, there would be only 51 cards to choose from, and there would be only three of whatever card you drew the first time. Drawing two cards without replacement are dependent events, because the result of the first draw changes the probability for the second draw.

If two cards are drawn at random from a standard deck, with replacement—that is, the first card is drawn, recorded, and replaced before the second card is drawn—the probability of drawing two aces is $\frac{4}{52} \times \frac{4}{52} = \frac{1}{13} \times \frac{1}{13} = \frac{1}{169} \approx 0.6\%$. If two cards are drawn without replacement, however, the probability of drawing two aces is $\frac{4}{52} \times \frac{3}{51} = \frac{1}{13} \times \frac{1}{17} = \frac{1}{221} \approx 0.5\%$.

EXERCISE 16·6

For questions 1 through 4, a single card is drawn at random from a standard deck.

1. What it the probability that the card is red and a king?

2. What is the probability that it is black and a face card?

3. What is the probability that it is a heart and shows a number between 2 and 5, inclusive?

4. What is the probability that it is a club and shows a number (2, 3, 4, 5, 6, 7, 8, 9, 10)?

For questions 5 through 10, two dice are rolled. One is red and one is white.

5. What is the probability of rolling a 6 on the red die and a 1 on the white die?

6. What is the probability of rolling a 2 on the red die and an odd number on the white die?

7. What is the probability of rolling a 4 on both dice?

8. What is the probability of rolling an odd number on the red die and an even number on the white die?

9. What is the probability that the total of the dice is 2?

10. What is the probability that the total of the dice is 7?

Compound probability: or

When you look at the probability of event A and event B happening, you multiply the probability of A times the probability of B. Multiplying these two fractions, both less than one, gives a smaller number. A smaller probability makes sense because it's harder to have a success if success is having two different things both happen than if success is having one thing happen.

What if you want to find the probability of getting a 5 or a 6 when you roll a die? If you're willing to accept either number as a success, your chance of success increases—in this case, it doubles. The probability of rolling a 5 or a 6 is the probability of rolling a 5 plus the probability of rolling a 6. $P(5 \text{ or } 6) = P(5) + P(6) = \frac{1}{6} + \frac{1}{6} = \frac{2}{6} = \frac{1}{3}$.

If you draw a card at random from a standard deck, what's the probability of drawing a king or an ace? The probability of drawing a king is $\frac{4}{52} = \frac{1}{13}$, and the probability of drawing an ace is also $\frac{4}{52} = \frac{1}{13}$, so the probability of drawing a king or an ace is $\frac{4}{52} + \frac{4}{52} = \frac{1}{13} + \frac{1}{13} = \frac{2}{13}$.

If you draw a card from a standard deck and you want to know the probability of drawing a king or a heart, start with the probability of drawing a king and the probability of a heart. $P(king) = \frac{4}{52} = \frac{1}{13}$ and $P(heart) = \frac{13}{52} = \frac{1}{4}$. Before you add, if you think for a minute, you'll realize there's a little problem: the king of hearts. The probability of a king was based on four kings, including the king of hearts. The probability of a heart was determined by also including the king of hearts. There's only one king of hearts, but it's being counted twice, so you'll have to make an adjustment for that.

$$P(\text{king or heart}) = P(\text{king}) + P(\text{heart}) - P(\text{king of hearts})$$
$$= \frac{4}{52} + \frac{13}{52} - \frac{1}{52}$$
$$= \frac{16}{52} = \frac{4}{13}$$

In general, the probability of A or B is the probability of A plus the probability of B minus the probability of A and B.

$$P(\text{A or B}) = P(\text{A}) + P(\text{B}) - P(\text{A and B})$$

You may wonder why the adjustment wasn't necessary in the earlier examples. The answer is that when you roll a die, getting a 5 and getting a 6 are mutually exclusive outcomes. There is no way that you can get a 5 and a 6 on the same roll. Drawing a king and drawing an ace are mutually exclusive. There is no card that is both a king and an ace. When event A and event B are mutually exclusive, the probability of A and B is zero, so $P(\text{A or B}) = P(\text{A}) + P(\text{B}) - P(\text{A and B})$ just becomes $P(\text{A or B}) = P(\text{A}) + P(\text{B})$.

EXERCISE
16·7

For questions 1 through 4, a single card is drawn at random from a standard deck of 52 cards. Decide if the events are mutually exclusive and find the probability.

1. Find the probability that the card is a 7 or an ace.

2. Find the probability that the card is a number less than 6 (2, 3, 4, or 5) or is a 10.

3. What is the probability that the card is a heart or a 7?

4. What is the probability that the card is black or a face card?

In questions 5 through 9, one marble is drawn at random from a bag that contains 24 marbles. Six of the marbles in the bag are red, 4 are blue, 3 are white, 9 are green, and 2 are yellow.

5. What is the probability that the chosen marble is red or blue?

6. What is the probability that the chosen marble is yellow or blue?

7. What is the probability that the chosen marble is green or white?

8. What is the probability that the chosen marble is yellow or red?

9. What is the probability that the chosen marble is red or orange?

Questions 10 through 15 ask you to use several of the rules you've learned. For each question, assume that a single card is drawn at random from a standard deck of 52 cards.

10. What is the probability that the card drawn is a 5 of hearts (5 and heart) or a 7 of spades (7 and spade)?

11. What is the probability that the card is a king or a queen and not a spade?

12. What is the probability that the card is not a red 9?

13. What is the probability that the card is not a 6 or a 7?

14. What is the probability that the card is black and not a face card?

15. What is the probability that the card is not an ace or a 10, and is a red card?

Conditional probability

We usually think that the probability of an event is fixed, unchanging, but actually, the probability may be different in different circumstances. If you draw two cards at random without replacement, the probability that the second card you draw is an ace depends on what card you drew first. If the first card you draw is an ace, the probability that the second will be an ace is $\frac{3}{51} = \frac{1}{17}$. If your first card is not an ace, the probability of an ace on the second draw is $\frac{4}{51}$.

If you are going to choose a student at random from the 55 women and 45 men who are studying math at a university and you have the information in the following table to tell you how those 100 people break down into graduate students and undergraduates, you can answer some questions about the probability that the student chosen will be a grad student or an undergrad, or the probability that the chosen student will be male or female.

	GRADUATE	UNDERGRADUATE	
Women	18	37	55
Men	17	28	45
	35	65	

The probability that a student chosen at random is a graduate student is $\frac{35}{100} = 35\%$. The probability that a student chosen at random is female is $\frac{55}{100} = 55\%$. But what if you choose someone at random *from the grad students*? What's the probability that you choose a woman? That's $\frac{18}{35} \approx 51.4\%$, because there are 18 women among the 35 grad students, and you know you're only looking at the grad students. The probability that you choose a woman if you're choosing from the whole pool is 55%, but the probability that the person you choose is a woman, given that the person is a grad student, is 51.4%. The probability of choosing a grad student from the entire pool is 35%, but the probability that you choose a grad student, given that your choice is a man, is 17 male grad students out of 45 men, or $\frac{17}{45} \approx 37.8\%$.

Questions 1 through 5 are based on the political registrations in a small town outlined in the following table.

	MALE	FEMALE	
Democrat	500	850	1,350
Republican	750	625	1,375
Independent	85	115	200
	1,335	1,590	2,925

1. What is the probability that a person chosen at random is a Republican, given that the person is male?

2. What is the probability that a person chosen at random is female, given that the person is a Republican?

3. What is the probability that a person chosen at random is an Independent, given that the person is female?

4. What is the probability that a person chosen at random is male, given that the person is a Republican?

5. What is the probability that a person chosen at random is female, given that the person is an Independent?

In questions 6 through 10, assume that two cards are drawn from a standard deck. The first card is drawn and not replaced, and then the second card is drawn.

6. What is the probability that the second card is an ace, given that the first card was an ace?

7. What is the probability that the second card is a black card, given that the first card was black?

8. What is the probability that the second card is a face card, given that the first card was a queen?

9. What is the probability that the second card is a face card, given that the first card was not?

10. What is the probability that the second card is red, given that the first card was red?

Statistics

Our lives are flooded with information, and the challenge for all of us is to make sense of it. Unless you can find patterns in the data, it's just a jumble of facts. One of the first ways you can find meaning is to organize the data into graphs, to give a general picture of what's going on. Then you can summarize the data by finding measures of center and spread, and look for ways to express the relationships you see.

Types of graphs

Line graphs are used to show the change in some quantity over time. The line graph in Figure 17-1 shows the change in the number of hot dogs sold at Jenny's Beach Bungalow over the course of a year. Each dot represents the sales for a particular month, but the lines connecting the points help to show the trend.

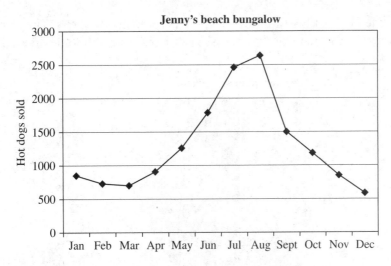

Figure 17-1

To investigate the relationship between two variables, form pairs of numbers in which the first number is the value of one variable and the second number is the corresponding value of the second variable. Then plot the points that represent those ordered pairs, and you have a graph called a **scatter plot** (Figure 17-2).

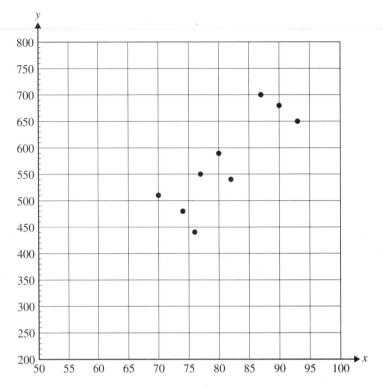

Figure 17-2

Circle graphs or pie charts illustrate how a whole is broken into parts. The circle graph in Figure 17-3 shows how the enrollment in arts courses at a particular school is divided among the various courses offered.

Enrollment in arts courses

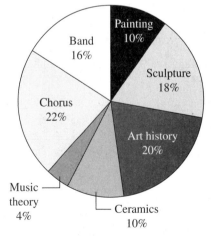

Figure 17-3

Bar graphs show the value of different quantities or of the same quantity at different times. The bar graph in Figure 17-4 compares the sales at a bookstore on the five business days of the week.

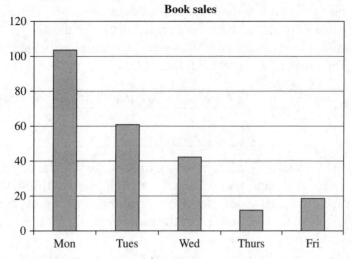

Figure 17-4

Histograms are variants of bar graphs that show the frequency with which each of the possible outcomes of an experiment occur (Figure 17-5). In a histogram, all the possible outcomes appear on the horizontal axis and the height of the bar tells you how often that outcome was observed. Because every possible outcome is covered, histograms have bars that touch.

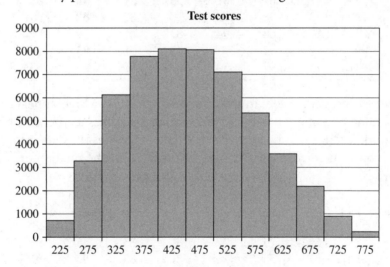

Figure 17-5

Types of data

Data comes in two forms: **quantitative** and **qualitative**. Quantitative data is numbers, and specifically, numbers with which it makes sense to do arithmetic. If George is 42 years old and Harry is 37 years old, subtracting their ages and saying that George is 5 years older than Harry makes sense. That data is quantitative. In contrast, George's zip code (02859) and Harry's zip code (11234) are numbers, but subtracting the zip codes wouldn't give any information. Although they're in the form of numbers, zip codes are actually qualitative data. Qualitative data is sometimes called categorical because it creates groups. A survey by a candy company that asks what new color should be added to an assortment will collect responses that forms groups of support for different colors.

Measures of center

The most common way to summarize data is to express the average result. What is meant by average varies according to the type of data under consideration. For qualitative or categorical data, the average is the value that occurs most frequently, called the **mode**.

A survey at a shopping mall asked 100 shoppers to name the principal item they were shopping for that day. The results are reported in the table.

CATEGORY	CLOTHING	SHOES	FURNITURE	APPLIANCES	SPORTING GOODS	ELECTRONICS
Number of responses	56	12	4	3	10	15

If you think of the table in its original form, as an unprocessed series of responses, you can imagine a list in which categories occur many times. It might look something like this: clothing, clothing, sporting goods, electronics, shoes, clothing, furniture, clothing, clothing . . . The mode is the most common response in that list, and from the more organized table, you can see that the most common response is clothing.

Although it's possible to find the mode of quantitative data also, the average of quantitative data is usually presented as the mean or the median. If the scores earned by members of the freshman class on a placement test are collected, those scores can be summarized by any one of the three types of averages.

PLACEMENT TEST SCORES

85	73	91	67	81	93	74	51	88	76
42	78	53	98	91	77	82	60	75	81
91	83	72	64	94	83	61	79	81	65
59	99	87	92	77	65	91	83	84	79
81	83	76	92	50	63	49	99	83	75

The mode, or most frequent value, in the list of scores is 83, which appears five times, but that most common value may not tell you as much about the scores as the median or the mean. The **median** is the middle value of any ordered set of data. It divides the data into two equal groups — half the data is above the median, and half is below. To find the median, it will be helpful to put the scores in order.

PLACEMENT TEST SCORES

42	59	65	74	77	81	83	84	91	93
49	60	65	75	77	81	83	85	91	94
50	61	67	75	78	81	83	87	91	98
51	63	72	76	79	81	83	88	92	99
53	64	73	76	79	82	83	91	92	99

If quantitative data is placed in order, either from smallest to largest or largest to smallest, the middle value is the median. If there is an even number of data points and therefore two values fall in the middle, the median is halfway between them.

The list of scores contains 50 values, so the median is midway between the twenty-fifth and twenty-sixth values. In this case, that means the median is midway between 79 and 81, so the median is 80.

The mode gives you a sense of what happens most often and the median gives you a sense of where the middle of the distribution is, but most people, when they talk about an average, think about the mean. The **mean** is the sum of all the data values, divided by the number of data values.

To find the mean of quantitative data, the individual data values are added up, and the total is divided by the number of values that have been added. If you again use the placement test score data, you need to first add up all 50 scores. (Of course, a calculator will be helpful.) You should find that the 50 scores total 3,856. Then divide 3,856 by 50—the number of scores—to find that the mean is 77.12.

Grouped data

Large sets of data are often organized into groups or classes that span several values. Classes should be of a uniform size, should cover all possible data values, and should be distinct. Every data value should fit in one and only one class.

When a large collection of data is organized into classes, it's easier to work with, but you sacrifice some information. If a survey asked a large number of people how many text messages they had sent in the previous 24 hours, the results might be grouped into classes as shown in the following table.

NUMBER OF TEXT MESSAGES SENT IN THE LAST 24 HOURS

CLASS	FREQUENCY
0–4	81
5–9	108
10–14	153
15–19	112
20–24	182
25–29	93
30–34	42

Recording the number of people who sent no texts, one text, two texts, and so on, would be unwieldy and would make it hard for you to see any patterns. On the other hand, you can't tell from this summary whether all 81 people in the first class said that they didn't text at all, or whether they gave different answers. You can't tell how many of the 42 people in the 30–34 class sent exactly 30 texts. Usually, you're willing to give up those specific bits of information in exchange for a clearer view of patterns.

To find the mean of grouped data, choose a class mark for each class. The **class mark** is a value that represents the group. It may be the highest value, the lowest value, or commonly, the midpoint of the class. Whatever you choose, be consistent for all the classes. In the text message example, you can use the midpoints of the classes as class marks. Multiply the frequency (the number of values in the class) by the class mark for each interval.

NUMBER OF TEXT MESSAGES SENT IN THE LAST 24 HOURS

CLASS	FREQUENCY	CLASS MARK	FREQUENCY × CLASS MARK
0–4	81	2	162
5–9	108	7	756
10–14	153	12	1,836
15–19	112	17	1,904
20–24	182	22	4,004
25–29	93	27	2,511
30–34	42	32	1,344
Total Frequency: 771			Total: 12,517

Then add the products (12,517), find the total number of responses (771), and divide 12,517 by 771 to find the mean. The mean for this survey was approximately 16.2 texts.

A group of 25 adults were asked to record the number of e-mails they received on one weekday. Their responses are shown in the table. Use this data for questions 1 through 3.

NUMBER OF E-MAILS RECEIVED IN A DAY				
284	63	73	83	45
932	74	28	92	65
153	317	46	75	93
24	382	28	49	101
182	109	28	33	16

1. What is the mode of the data?

2. What is the median of the data?

3. What is the mean of the data?

4. A candy company asked visitors to their website to vote on the flavor of candy they should introduce. The results are shown in the table. What is the mode?

Flavor	Peanut butter	Chocolate fudge	Peppermint twist	Almond crunch	Red raspberry
Number of votes	47,392,018	938,764,402	38,193,810	3,810,437	3,920,147

Use the data in the following table to answer questions 5 through 7.

AGE OF WOMEN AT THE TIME OF MARRIAGE						
17	22	24	27	28	31	34
19	22	25	27	28	31	35
19	23	26	27	29	32	37
20	23	26	27	29	32	40
21	23	27	28	30	33	42

5. What marriage age was most commonly reported by the women surveyed?

6. Find the median age at the time of marriage.

7. What is the mean age of marriage?

Use the data in this table to answer questions 8 through 10.

NUMBER OF INNINGS PITCHED BY YANKEES PITCHERS AS OF JUNE 15, 2011		
GREATER THAN	LESS THAN OR EQUAL TO	# OF PITCHERS
0	10	4
10	20	5
20	30	4
30	40	0
40	50	0

NUMBER OF INNINGS PITCHED BY YANKEES PITCHERS AS OF JUNE 15, 2011		
GREATER THAN	LESS THAN OR EQUAL TO	# OF PITCHERS
50	60	0
60	70	2
70	80	1
80	90	1
90	100	0
100	110	1
	Total frequency:	18

(Source: Major League Baseball, http://newyork.yankees.mlb.com/stats/
sortable_player_stats.jsp?c_id=nyy)

8. What is the mode of the innings pitched data?

9. What is the median number of innings pitched?

10. Find the mean number of innings pitched.

The following table shows the annual salaries for employees at a small company. The highest salary is that of the CEO of the company.

ANNUAL SALARIES AT ZYX WIDGETS			
$8,500	$18,500	$30,000	$50,000
$10,000	$20,000	$40,000	$80,000
$12,000	$25,000	$45,000	$120,000

11. Find the median salary at ZYX Widgets.

12. Find the mean salary at ZYX Widgets.

For questions 13 through 15, assume that the CEO of ZYX Widgets doubles her own salary, from $120,000 to $240,000.

13. What is the median of the revised salary data?

14. What is the mean of the revised salary data?

15. Why does doubling the CEO's salary have a greater effect on the mean than the median?

The CEO of ZYX Widgets has a change of heart and decides not to double her own salary. Instead, she decides to distribute those funds by giving every employee, including herself, a raise of $10,000. Use this new salary data for questions 16 through 18.

REVISED ANNUAL SALARIES AT ZYX WIDGETS			
$18,500	$28,500	$40,000	$60,000
$20,000	$30,000	$50,000	$90,000
$22,000	$35,000	$55,000	$130,000

16. Find the median of the revised salaries at ZYX Widgets.

17. Find the mean of the revised salaries at ZYX Widgets.

18. Predict, without further calculation, what the median and mean salaries will be if, the following year, the CEO gives everyone an additional $5,000 raise.

19. Chet's mean score for three rounds of golf was 90. If he recorded scores of 90 and 92 for two of the rounds, what was his score for the third round?

20. After the first four tests of the semester, April has a mean score of 88. What score must she earn on the fifth test to raise her average to 90?

Measures of spread

While the mean, median, and mode tell you about the center of the data, where the data clusters, it's sometimes useful to know how spread out the data is. If 10 people take a test and you know the mean score is 80%, you have some idea of how people performed, but you may not have the whole picture. Have a look at the two sets of test scores in the table below.

| Test 1 | 84% | 83% | 82% | 81% | 80% | 80% | 79% | 78% | 77% | 76% | Mean: 80% Median: 80% |
| Test 2 | 100% | 100% | 100% | 100% | 100% | 60% | 60% | 60% | 60% | 60% | Mean: 80% Median: 80% |

Both tests have a mean of 80%, but on the first test, everyone scored within a few points of the mean. On the second test, half the group scored much higher than the mean and the other half, much lower. When there's too much data to just look at a list, having some indication of how spread out the data is can help you understand what's going on.

One way to get a sense of the spread of the data is to find the difference between the highest and the lowest value. That difference is called the range. In test 1, the range is $84\% - 76\% = 8\%$, while in test 2 the range is $100\% - 60\% = 40\%$. Knowing that one test has a range of 8% and the other a range of 40% immediately tells you that the results, despite having the same mean and median, were very different on the two tests.

Another way to get a picture of the distribution of the data is to break it into groups. You know that the median divides the top 50% from the bottom 50%. If you found the median of the lower half and the median of the upper half, you would divide the data into four groups of equal size. These groups are called quartiles. If you divided the data into five groups, you'd have quintiles. Six groups would be sextiles, and so on, but the most common breakdowns are **quartiles** and **percentiles**.

Percentiles break the data into 100 groups of equal size. If a data value is "at the 80th percentile," that means that 80% of the data is below it, and 20% above. The median value is the 50th percentile because half the data is below it. In the test data above, the 80th percentile on the first test is 82.5%, but for the second test the 80th percentile is 100%. Eight scores are below those, and two are above. If you think about the spread as moving away from the center, or 50th percentile, when you've moved up to the 80th percentile on the first test, you've only moved to 82.5%, but on the second test, you've moved to a much higher score. The second set of test scores are much more spread out.

When the data is divided into four sections, or quartiles, the spread is sometimes described by the interquartile range. The **interquartile range**, or IQR, is the difference between the value that divides the first and second quartile, called Q1, and the value that divides the third and fourth quartile, called Q3. You can think of it as the range of the middle 50% of the values. For the first test, Q1 = 78, Q3 = 82, and IQR = Q3 – Q1 = 82 – 78 = 4. For the second test, Q1 = 60, Q3 = 100, and IQR = Q3 – Q1 = 100 – 60 = 40. Clearly, the second test has a much greater spread.

Another common measure of the spread of data is the standard deviation. The calculation of the standard deviation takes several steps, and you'll definitely want to use a calculator, but the important thing to remember is that the bigger the standard deviation, the more spread out the data is.

To find a standard deviation, first find the mean, and then subtract the mean from each data value. You'll get positive numbers for values above the mean and negative numbers for values below the mean, but what you're really interested in is how far from the mean each value is, so square each result, which makes everything positive, and add up the squares. Divide by one less than the number of data values, and then take the square root of the result. (That last square root will undo the earlier squaring.)

The demonstration below uses a small set of data, with x representing the data value and μ the symbol for the mean. The mean for this data is 12.

x	$x - \mu$	$(x - \mu)^2$
4	−8	64
8	−4	16
13	1	1
15	3	9
20	8	64

The total of the $(x - \mu)^2$ column is 154. There are five values, so divide the total by one less than that. $154 \div 4 = 38.5$. Then take the square root. $\sqrt{38.5} \approx 6.2$. The standard deviation is approximately 6.2.

You can probably see why calculators and computers are helpful when finding the standard deviation of a set of data. You probably wouldn't be asked to find the standard deviation by hand for a data set of more than a few numbers, but you will use information about the standard deviation to make decisions about data. Suppose you scored an 82 on a test with a mean of 75. Did you do well? Certainly, your score is above average; it's higher than the mean. When you add some information about the standard deviation, however, you get a better picture. If the mean was 75 and the standard deviation was 3, scores were tightly clustered around the mean, and yours was more than two standard deviations above the mean. You're out in front of the pack. On the other hand, if the mean was 75 and the standard deviation was 12, there was a wider spread and your score falls within the first standard deviation. Your score is still above average, but not as much of a standout.

EXERCISE
17·2

Use the placement test score data in the following table to answer questions 1 through 3.

PLACEMENT TEST SCORES									
42	59	65	74	77	81	83	84	91	93
49	60	65	75	77	81	83	85	91	94
50	61	67	75	78	81	83	87	91	98
51	63	72	76	79	81	83	88	92	99
53	64	73	76	79	82	83	91	92	99

1. What is the range of the placement test scores?

2. The median of the placement test scores is 80. What are the values of Q1 and Q3?

3. What is the interquartile range?

Use the data on age at time of marriage to answer questions 4 through 6.

AGE OF WOMEN AT THE TIME OF MARRIAGE						
17	22	24	27	28	31	34
19	22	25	27	28	31	35
19	23	26	27	29	32	37
20	23	26	27	29	32	40
21	23	27	28	30	33	42

4. The median of the age of marriage data for women is 27. Find the value of Q1 and the value of Q3.

5. What is the interquartile range for the age of marriage data for women?

6. What is the range of the age of marriage data for women?

The following table shows the age of men at the time of marriage. Use this data for questions 7 through 10.

AGE OF MEN AT THE TIME OF MARRIAGE						
19	24	28	30	33	35	41
21	25	28	31	33	36	45
23	25	28	31	33	36	48
23	26	28	32	34	37	49
24	27	28	32	34	39	52

7. Find the range of the age of marriage data for men.

8. Find the median of the age of marriage data for men.

9. Find the values of Q1 and Q3 of the age of marriage data for men.

10. Find the interquartile range of the age of marriage data for men.

11. Based on your answers to questions 4 through 10, compare the data on marriage ages of women with the data for men. Which has the greater spread?

The mean of the age of marriage data for women is 27.5, with a standard deviation of 5.8, while the mean of the age of marriage data for men is 31.9, with a standard deviation of 7.9.

12. Which would be a more significant report: a woman who reported her age at time of marriage as 45 or a man who reported his age at time of marriage as 45?

Use the data on revised annual salaries at ZYX Widgets to answer questions 13 through 15.

REVISED ANNUAL SALARIES AT ZYX WIDGETS			
$20,000	$30,000	$50,000	$90,000
$22,000	$35,000	$55,000	$130,000

13. The mean salary at ZYX Widgets is $54,000. Find the standard deviation.

14. Find the values of Q1 and Q3, and the interquartile range.

15. Find the range of salaries.

The data in the table shows the number of e-mails received in one day by a sample of adults. Use this data to answer questions 16 through 19.

NUMBER OF E-MAILS RECEIVED IN A DAY				
284	63	73	83	45
932	74	28	92	65
153	317	46	74	93
24	382	28	49	101
182	109	29	33	16

16. Find the range of the number of e-mails received.

17. The data on e-mails received has a median of 74. Find the value of Q1 and the value of Q3.

18. Find the interquartile range.

19. The median of the data is 74, but the mean is 135. Why is the mean so much larger than the median?

20. Which would you consider a better performance: a score of 85 on a test with a mean of 75 and a standard deviation of 10, or a score of 85 on a test with a mean of 78 and a standard deviation of 3?

Expressing relationships

Does aluminum in antiperspirants cause cancer? Does sugar make children hyperactive? Often the reason to collect data is the desire to understand connections between events. If you take vitamin C, will you avoid the common cold? To find out, you'd need to look at people who received different amounts of vitamin C and note how many colds they had. You'd want to compare that information to information about people who didn't take any extra vitamin C, and how many colds they had. In order to make sense of that information you'd need to express the relationship between vitamin C intake and colds.

Common relationships

The first and most fundamental question you'll want to answer is, when one factor increases, does the other factor increase or decrease? If you take more vitamin C, do you get more colds or fewer colds? (Obviously, if there's no change, that tells you something too.) If you run more miles in training, does your race time go up or down? Do students who get good grades in their algebra class score high on the math section of the SAT?

To answer that last question, you might look at the algebra grades and SAT scores of students. If you look at the information, arranged in increasing order of algebra grade, you'll see that, while the relationship certainly isn't perfect, the students with higher algebra scores tend to have higher SAT scores.

ALGEBRA GRADE	70	74	76	77	80	82	87	90	93
MATH SAT	510	480	440	550	590	540	700	680	650

Once you've established that basic relationship, you may want to try to classify the pattern further. Making a graph of the data can help you identify that pattern. If you place algebra grades on the horizontal axis and SAT scores on the vertical axis, each student's score is represented by a single point. The graph that results is called a **scatter plot** (Figure 17-6). The shape of that scatter plot can help you specify the relationship.

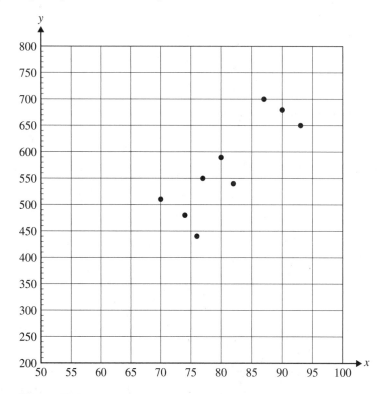

Figure 17-6

The primary question you want to ask is whether the points suggest a line or a curve. If, like the scatter plot in Figure 17-6, the arrangement of the points suggests a line, like the function in Figure 17-7a, you can say the relationship is linear. If the scatter plot seems to curve, the relationship may be a power function, like the graph in Figure 17-7b, if the curve is gentle, or an exponential function, like Figure 17-7c, if it's flat on one end and steep on another.

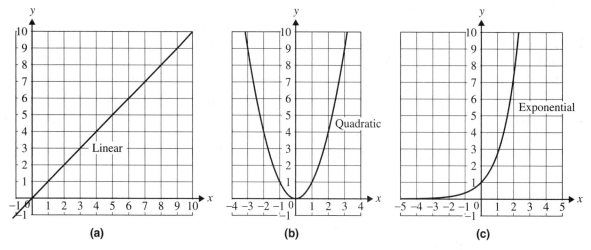

Figure 17-7

Finding the precise equations that describe those relationships is a complicated process, but it is possible to find the equation of a line that reasonably represents the data. This trend line won't hit every point, but it will follow the direction of the scatter. As you look at the scatter plot, try to locate a line that follows the direction of the points and cuts through the middle of the scatter so that there are approximately as many points above the line as below.

To find the equation of such a trend line, you'll need two points. Choose any two points that fall on your trend line. You'll probably find two data points that fall on the line, but if you don't, just use any two points on the line. In the example in Figure 17-6, the trend line passes through the points (74, 480) and (90, 680). Find the slope of the line using the slope formula $m = \dfrac{y_2 - y_1}{x_2 - x_1}$.

Here, $m = \dfrac{680 - 480}{90 - 74} = \dfrac{200}{16} = 12.5$. We have already determined that the trend is linear, so plug the slope and the coordinates of one of your points into the point-slope form of a linear equation:

$y - y_1 = m(x - x_1)$. The equation for the trend line in Figure 17-6 is

$$y - 680 = 12.5(x - 90)$$
$$y - 680 = 12.5x - 1125$$
$$y = 12.5x - 445$$

EXERCISE
17·3

For the data presented in questions 1 through 5, decide which of the following statements is true.

A. As the first variable increases, the second increases.

B. As the first variable increases, the second decreases.

C. As the first variable increases, the second first increases then decreases.

D. As the first variable increases, the second first decreases then increases.

E. There is no apparent connection between the behaviors of the two variables.

1.

x	0	1	2	3	4	5	6	7	8
y	7	2	−1	−2	−1	2	7	14	23

2.

a	10	20	30	40	50	60	70	80	90
b	720	690	660	630	600	570	540	510	480

3.

t	0.1	0.4	0.6	0.9	1.3	1.7	2.2	2.5	2.9
v	1.105	1.492	1.822	2.460	3.669	5.474	9.025	12.182	18.174

4.

x	267	283	297	312	329	337	341	358	379
y	246.5	270.5	291.5	314	339.5	351.5	357.5	383	414.5

5.

t	0.5	0.9	1.7	2.1	2.8	3.4	4.9	5.7	6.5
f	3.903	7.575	13.287	15.327	17.588	18.2	14.375	9.207	1.863

For questions 6 through 10, examine the scatter plot and tell whether the relationship between the variables could best be described as linear, quadratic, exponential, or other.

6.

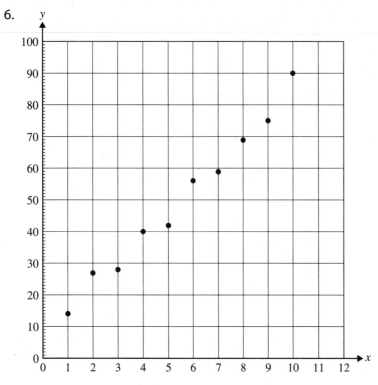

7.

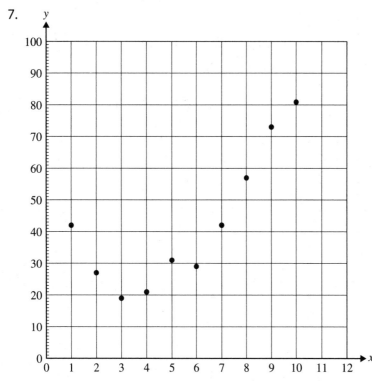

8.

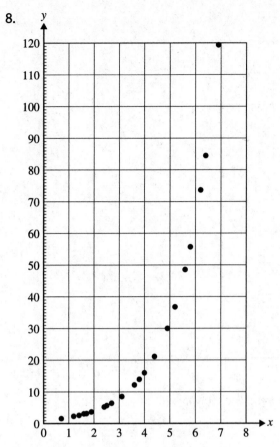

9.

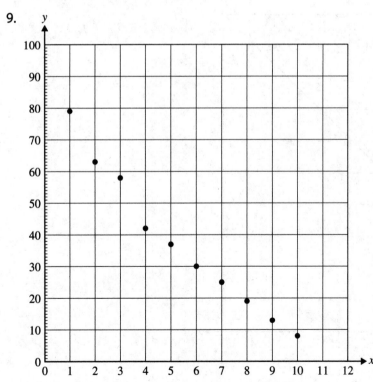

10.

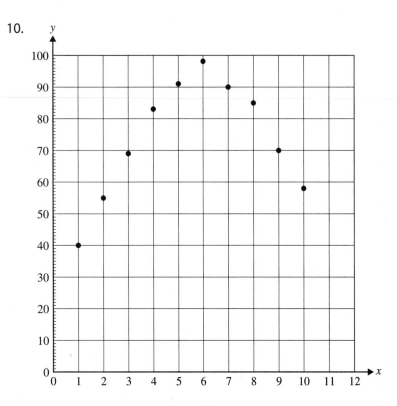

For questions 11 through 15, find the equation of the trend line shown.

11.

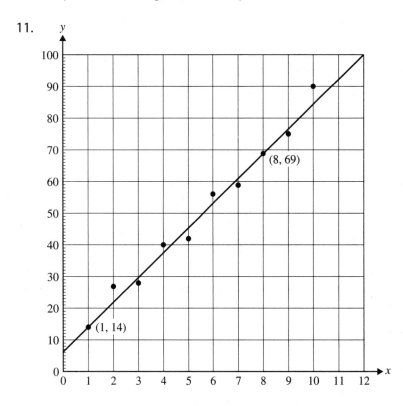

(8, 69)

(1, 14)

12.

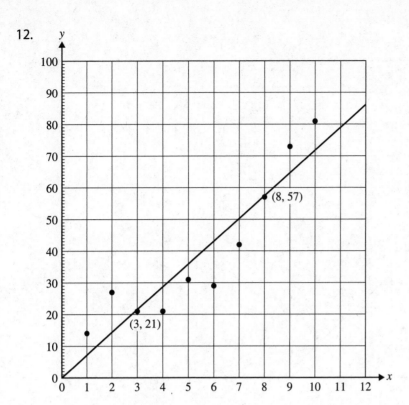

13.

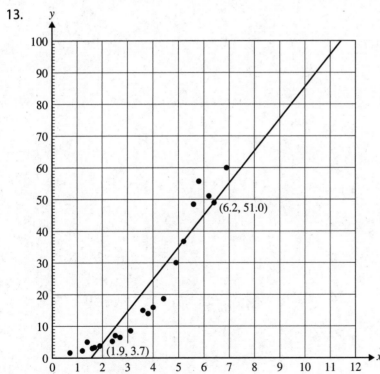

14.

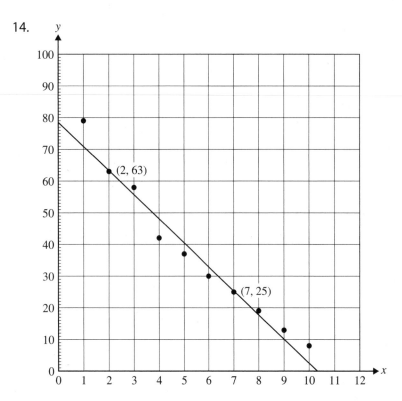

15.

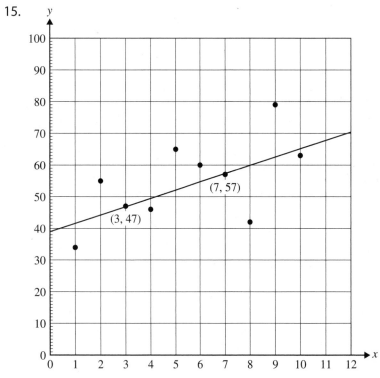

For questions 16 through 20, draw a trend line and find its equation.

16.

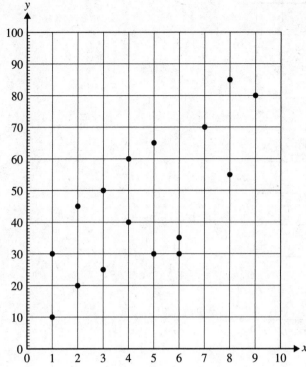

17.

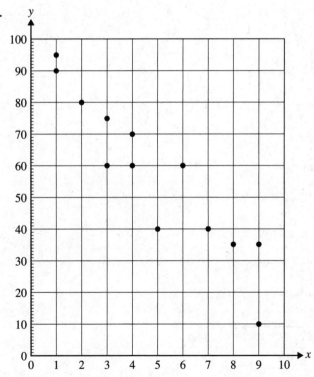

18.

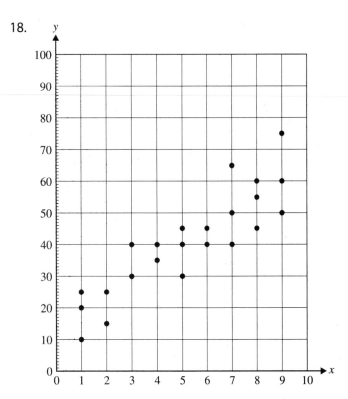

19.

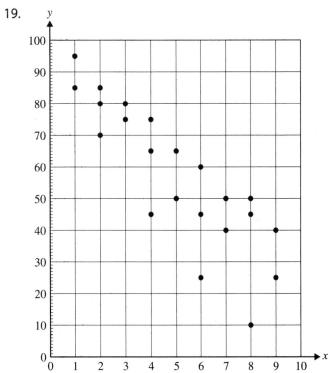

20.

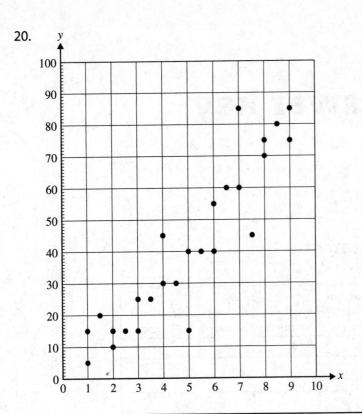

Answer key

1 Numbers and arithmetic

1·1

		Natural or Counting Numbers	Whole Numbers	Rational Numbers	Irrational Numbers
1.	7	X	X	X	
2.	$\dfrac{3}{4}$			X	
3.	1.9			X	
4.	0		X	X	
5.	$4\dfrac{5}{7}$			X	
6.	193	X	X	X	
7.	18.328			X	
8.	18.3284502193…				X
9.	$6.\overline{29}$			X	
10.	$\dfrac{11}{23}$			X	

1·2

1. To say "4.38 is a real number," you could write $4.38 \in \mathbb{R}$.
2. If the symbol for the irrational numbers is $\mathbb{Q}$, you can say "the irrational numbers are a subset of the real numbers" by writing $\mathbb{Q} \subset \mathbb{R}$.
3. $6 \times 4 = 24$
4. $6 + 4 = 10$
5. $6 - 4 = 2$
6. $6 \div 4 = 1.5$
7. $\dfrac{16}{3} \in$ the rational numbers or the real numbers
8. $3 \times 3 \times 3 \times 3 =$ three to the fourth power
9. $\sqrt{9} > \sqrt[3]{8}$ or $\sqrt{9} \neq \sqrt[3]{8}$. ($\sqrt{9} = 3$ but $\sqrt[3]{8} = 2$)
10. $7^2 < 5 \times 10$ or $7^2 \neq 5 \times 10$
11. The product of 8 and 3 = the quotient of 96 and 4.
12. The sum of 18 and 12 > the square of 4.
13. The difference of 100 and 80 < 5 squared.
14. Two to the third power < three squared.
15. The quotient of 84 and 6 = the product of 7 and 2.
16. True: 57 is a multiple of 3.
17. False: 483 is NOT a multiple of 9. $4 + 8 + 3 = 15$, which is not a multiple of 9.
18. False: 452 is a multiple of 6. 452 is even but $4 + 5 + 2 = 11$, which is not a multiple of 3.
19. True: 6^2 is a multiple of 9. $6^2 = 36$ and $3 + 6 = 9$, so 36 is a multiple of 9.

20. True: $48 \div 4$ is a multiple of 6. $48 \div 4 = 12$, which is both even and a multiple of 3.

21. True: 2,012 is a multiple of 4. The last 2 digits form the number 12, which is a multiple of 4.

23. False: 51 is not prime. $5 + 1 = 6$, so 51 is divisible by 3 (and 17) as well as itself and one.

24. True: $68 \div 4$ is prime. $68 \div 4 = 17$, which is prime.

25. True: $\sqrt{10}$ is a natural number. $\sqrt{10}$ is between 3 and 4.

1·3

1. $35 - 17 + 22 = 18 + 22$
$= 40$

2. $505 - 100 \times 3 = 505 - 300$
$= 205$

3. $125 - 18 \div 6 \times 21 + 7 = 125 - 3 \times 21 + 7$
$= 125 - 63 + 7$
$= 62 + 7$
$= 69$

4. $13 \times 20 + 35 + (48 - 15) \div 11 = 13 \times 20 + 35 + 33 \div 11$
$= 260 + 35 + 33 \div 11$
$= 260 + 35 + 3$
$= 295 + 3$
$= 298$

5. $20 \times 3 \div 4 + 23 - 17 + 83 = 60 \div 4 + 23 - 17 + 83$
$= 15 + 23 - 17 + 83$
$= 38 - 17 + 83$
$= 21 + 83$
$= 104$

6. $234 \div 6 \times 2 + 103 = 39 \times 2 + 103$
$= 78 + 103$
$= 181$

7. $(200 + 300 \times 25 \div 15) \div 10 = (200 + 7{,}500 \div 15) \div 10$
$= (200 + 500) \div 10$
$= 700 \div 10$
$= 70$

8. $2 \times 3 + 5 - 25 \times 4 \div 10 = 6 + 5 - 25 \times 4 \div 10$
$= 6 + 5 - 100 \div 10$
$= 6 + 5 - 10$
$= 11 - 10$
$= 1$

9. $2 \times (3 + 5) - 25 \times 4 \div 10 = 2 \times 8 - 25 \times 4 \div 10$
$= 16 - 25 \times 4 \div 10$
$= 16 - 100 \div 10$
$= 16 - 10$
$= 6$

10. $5^2 \div (12 - 7) + 8 \times 3^2 = 5^2 \div 5 + 8 \times 3^2$
$= 25 \div 5 + 8 \times 9$
$= 5 + 72$
$= 77$

1·4

1. $|-4| = 4$
2. $|+3| = 3$
3. $|0| = 0$
4. $|178| = 178$
5. $|-192| = 192$
6. $-8 + 9 = 1$
7. $3 + -5 = -2$
8. $-7 + 11 = 4$
9. $12 + -9 = 3$
10. $-19 + 25 = 6$
11. $33 + -47 = -14$
12. $-87 + -91 = -178$
13. $44 + 35 = 79$
14. $-101 + 110 = 9$
15. $93 + -104 = -11$

1·5

1. $19 - (-8) = 19 + 8 = 27$
2. $-27 - 14 = -27 + -14 = -41$
3. $-35 - (-29) = -35 + 29 = -6$
4. $42 - (-37) = 42 + 37 = 79$
5. $-56 - 44 = -56 + -44 = -100$
6. $62 - 28 = 62 + -28 = 34$
7. $71 - (-35) = 71 + 35 = 106$
8. $-84 - 70 = -84 + -70 = -154$
9. $-90 - (-87) = -90 + 87 = -3$
10. $107 - 112 = 107 + -112 = -5$

1·6

1. $18 \times 2 = 36$
2. $-5 \div -1 = 5$
3. $-7 \times 3 = -21$
4. $50 \div -5 = -10$
5. $6 \times -4 = -24$
6. $-12 \div -3 = 4$
7. $-13 \times -5 = 65$
8. $49 \div -7 = -7$
9. $11 \times -3 = -33$
10. $-20 \div 4 = -5$
11. $-34 \times 3 = -102$
12. $110 \div -55 = -2$
13. $9 \times -8 = -72$
14. $-42 \div 7 = -6$
15. $-6 \times 12 = -72$
16. $(-3)^2 = 9$
17. $2^3 = 8$
18. $(-2)^3 = -8$
19. $-5^2 = -25$
20. $(-1)^4 = 1$

2   Factors and primes

2·1
1. 17 is prime.
2. 29 is prime.
3. 111 is composite (3 × 37).
4. 239 is prime.
5. 517 is composite (11 × 47).
6. 613 is prime.
7. 852 is composite (2 × 426).
8. 741 is composite (3 × 247).
9. 912 is composite (2 × 456).
10. 2,209 is composite (47 × 47).

2·2
1. $48 = 2^4 \times 3$
2. $92 = 2^2 \times 23$
3. $120 = 2^3 \times 3 \times 5$
4. $168 = 2^3 \times 3 \times 7$
5. $143 = 11 \times 13$
6. $300 = 2^2 \times 3 \times 5^2$
7. $189 = 3^3 \times 7$
8. $364 = 2^2 \times 7 \times 13$
9. $1,944 = 2^3 \times 3^5$
10. $5,225 = 5^2 \times 11 \times 19$

2·3
1. $a + b$ is even.
2. $x + y$ is even.
3. $a + x$ is odd.
4. $a \cdot b$ is even, so $a \cdot b + y$ is odd.
5. $a + x$ is odd, so $y(a + x)$ is odd.
6. $a \cdot x$ is even and $b \cdot y$ is even, so $a \cdot x + b \cdot y$ is even.
7. $a - b$ is even.
8. $x - y$ is even.
9. $a \cdot y$ is even, $b \cdot x$ is even, and $a \cdot y - b \cdot x$ is even.
10. a^2 is even, $b \cdot y$ is even, and x^2 is odd. $a^2 + b \cdot y$ is even but $a^2 + b \cdot y - x^2$ is odd.

2·4
1. False: 9 is NOT a factor of 89. $8 + 9 = 17$, which is not divisible by 9.
2. True: 6 is a factor of 84. 84 is divisible by 2 and $8 + 4 = 12$, so 84 is divisible by 3.
3. False: 4 is NOT a factor of 8,710. 10 is not divisible by 4.
4. True: 5 is a factor of 790. 790 ends in 0.
5. True: 11 is a factor of 209. There is no helpful test. Just divide. $209 \div 11 = 19$.
6. $67 = 11 \times 6 + 1$
7. $83 = 7 \times 11 + 6$
8. $194 = 13 \times 14 + 12$
9. $1,197 = 3 \times 399 + 0$
10. $401 = 8 \times 50 + 1$

2·5
1. The GCF of 35 and 42 is 7.
2. The GCF of 48 and 60 is 12.
3. The GCF of 64 and 96 is 32.
4. The GCF of 65 and 117 is 13.
5. The GCF of 93 and 98 is 1. 93 and 98 are relatively prime.
6. The GCF of 98 and 105 is 7.
7. The GCF of 105 and 495 is 15.
8. The GCF of 121 and 169 is 1. 121 and 169 are relatively prime.
9. The GCF of 144 and 342 is 18.
10. The GCF of 275 and 2,125 is 25.

2·6
1. The LCM of 14 and 24 is 168.
2. The LCM of 34 and 85 is 170.
3. The LCM of 33 and 55 is 165.
4. The LCM of 90 and 72 is 360.
5. The LCM of 36 and 27 is 108.
6. The LCM of 35 and 42 is 210.
7. The LCM of 88 and 110 is 440.
8. The LCM of 48 and 60 is 240.
9. The LCM of 175 and 140 is 700.
10. The LCM of 64 and 96 is 192.

3   Fractions

3·1
1. $\dfrac{45}{60} = \dfrac{3}{4}$
2. $\dfrac{27}{36} = \dfrac{3}{4}$
3. $\dfrac{42}{126} = \dfrac{1}{3}$
4. $\dfrac{28}{49} = \dfrac{4}{7}$
5. $\dfrac{75}{125} = \dfrac{3}{5}$
6. $\dfrac{189}{369} = \dfrac{21}{41}$
7. $\dfrac{88}{110} = \dfrac{4}{5}$
8. $\dfrac{35}{63} = \dfrac{5}{9}$
9. $\dfrac{105}{280} = \dfrac{3}{8}$

10. $\dfrac{153}{187} = \dfrac{9}{11}$

11. $\dfrac{5}{6} = \dfrac{15}{18}$

12. $\dfrac{4}{7} = \dfrac{20}{35}$

13. $\dfrac{2}{9} = \dfrac{12}{54}$

14. $\dfrac{3}{14} = \dfrac{18}{84}$

15. $\dfrac{7}{10} = \dfrac{77}{110}$

16. $\dfrac{1}{2} = \dfrac{12}{24}$

17. $\dfrac{5}{16} = \dfrac{30}{96}$

18. $\dfrac{7}{8} = \dfrac{42}{48}$

19. $\dfrac{15}{16} = \dfrac{75}{80}$

20. $\dfrac{11}{12} = \dfrac{121}{132}$

3·2

1. $\dfrac{6}{7} \times \dfrac{1}{3} = \dfrac{6}{21} = \dfrac{2}{7}$

2. $\dfrac{12}{17} \times \dfrac{1}{6} = \dfrac{12}{102} = \dfrac{2}{17}$

3. $\dfrac{15}{23} \times \dfrac{1}{5} = \dfrac{15}{115} = \dfrac{3}{23}$

4. $\dfrac{24}{35} \times \dfrac{1}{8} = \dfrac{24}{280} = \dfrac{3}{35}$

5. $\dfrac{9}{10} \times \dfrac{1}{3} = \dfrac{9}{30} = \dfrac{3}{10}$

6. $\dfrac{4}{5} \times \dfrac{1}{2} = \dfrac{4}{10} = \dfrac{2}{5}$

7. $\dfrac{14}{25} \times \dfrac{1}{7} = \dfrac{14}{175} = \dfrac{2}{25}$

8. $\dfrac{6}{7} \times \dfrac{1}{6} = \dfrac{6}{42} = \dfrac{1}{7}$

9. $\dfrac{39}{50} \times \dfrac{1}{13} = \dfrac{39}{650} = \dfrac{3}{50}$

10. $\dfrac{49}{100} \times \dfrac{1}{7} = \dfrac{49}{700} = \dfrac{7}{100}$

3·3

1. $\dfrac{3}{\cancel{6}} \times \dfrac{\cancel{6}}{7} = \dfrac{3}{7}$

2. $\dfrac{\cancel{2}}{\underset{3}{9}} \times \dfrac{3}{\underset{4}{\cancel{8}}} = \dfrac{1}{12}$

3. $\dfrac{\cancel{7}}{\underset{2}{12}} \times \dfrac{6}{\cancel{7}} = \dfrac{1}{2}$

4. $\dfrac{\overset{3}{\cancel{9}}}{11} \times \dfrac{2}{\cancel{3}} = \dfrac{6}{11}$

5. $\dfrac{\cancel{2}}{7} \times \dfrac{\cancel{7}}{\underset{6}{12}} = \dfrac{1}{6}$

6. $\dfrac{\cancel{2}}{3} \times \dfrac{\overset{5}{\cancel{15}}}{\underset{11}{\cancel{22}}} = \dfrac{5}{11}$

7. $\dfrac{3}{4} \times \dfrac{5}{\underset{7}{\cancel{21}}} = \dfrac{5}{28}$

8. $\dfrac{3}{\cancel{5}} \times \dfrac{\overset{2}{\cancel{10}}}{11} = \dfrac{6}{11}$

9. $\dfrac{\overset{3}{\cancel{9}}}{\underset{5}{10}} \times \dfrac{2}{\cancel{3}} = \dfrac{3}{5}$

10. $\dfrac{1}{\underset{2}{10}} \times \dfrac{\cancel{5}}{7} = \dfrac{1}{14}$

3·4

1. $\dfrac{3}{7} \times \dfrac{5}{8} = \dfrac{15}{56}$

2. $\dfrac{1}{4} \times \dfrac{5}{7} = \dfrac{5}{28}$

3. $\dfrac{\overset{1}{\cancel{2}}}{3} \times \dfrac{5}{\underset{4}{\cancel{8}}} = \dfrac{5}{12}$

4. $\dfrac{1}{\underset{1}{\cancel{5}}} \times \dfrac{\overset{2}{\cancel{10}}}{11} = \dfrac{2}{11}$

5. $\dfrac{\overset{3}{\cancel{12}}}{13} \times \dfrac{3}{\underset{1}{\cancel{4}}} = \dfrac{9}{13}$

6. $\dfrac{\overset{1}{\cancel{5}}}{\underset{1}{8}} \times \dfrac{\overset{2}{\cancel{16}}}{\underset{5}{\cancel{25}}} = \dfrac{2}{5}$

7. $\dfrac{\overset{1}{\cancel{7}}}{\underset{2}{10}} \times \dfrac{\overset{1}{\cancel{5}}}{\underset{2}{\cancel{14}}} = \dfrac{1}{4}$

8. $\dfrac{\overset{1}{\cancel{8}}}{\underset{1}{11}} \times \dfrac{\overset{3}{\cancel{33}}}{\underset{8}{\cancel{64}}} = \dfrac{3}{8}$

9. $\dfrac{\overset{3}{\cancel{6}}}{\underset{1}{\cancel{7}}} \times \dfrac{\overset{7}{\cancel{49}}}{\underset{25}{\cancel{50}}} = \dfrac{21}{25}$

10. $\dfrac{\overset{2}{\cancel{24}}}{\underset{5}{\cancel{25}}} \times \dfrac{\overset{7}{\cancel{35}}}{\underset{3}{\cancel{36}}} = \dfrac{14}{15}$

3·5

1. $\frac{3}{5} \div \frac{1}{2} = \frac{3}{5} \times \frac{2}{1} = \frac{6}{5} = 1\frac{1}{5}$

2. $\frac{5}{7} \div \frac{1}{5} = \frac{5}{7} \times \frac{5}{1} = \frac{25}{7} = 3\frac{4}{7}$

3. $\frac{9}{10} \div \frac{3}{5} = \frac{\cancel{9}^{3}}{\cancel{10}_{2}} \times \frac{\cancel{5}^{1}}{\cancel{3}_{1}} = \frac{3}{2} = 1\frac{1}{2}$

4. $\frac{15}{28} \div \frac{35}{48} = \frac{\cancel{15}^{3}}{\cancel{28}_{7}} \times \frac{\cancel{48}^{12}}{\cancel{35}_{7}} = \frac{36}{49}$

5. $\frac{6}{7} \div \frac{3}{14} = \frac{\cancel{6}^{2}}{\cancel{7}} \cdot \frac{\cancel{14}^{2}}{\cancel{3}} = \frac{4}{1} = 4$

6. $\frac{8}{33} \div \frac{4}{11} = \frac{\cancel{8}^{2}}{\cancel{33}_{3}} \cdot \frac{\cancel{11}}{\cancel{4}} = \frac{2}{3}$

7. $\frac{5}{21} \div \frac{11}{14} = \frac{5}{\cancel{21}_{3}} \cdot \frac{\cancel{14}^{2}}{11} = \frac{10}{33}$

8. $\frac{8}{35} \div \frac{4}{15} = \frac{\cancel{8}^{2}}{\cancel{35}_{7}} \cdot \frac{\cancel{15}^{3}}{\cancel{4}} = \frac{6}{7}$

9. $\frac{2}{7} \div \frac{4}{5} = \frac{\cancel{2}}{7} \cdot \frac{5}{\cancel{4}_{2}} = \frac{5}{14}$

10. $\frac{25}{52} \div \frac{20}{39} = \frac{\cancel{25}^{5}}{\cancel{52}_{4}} \cdot \frac{\cancel{39}^{3}}{\cancel{20}_{4}} = \frac{15}{16}$

3·6

1. $\frac{4}{7} + \frac{1}{7} = \frac{5}{7}$

2. $\frac{6}{11} - \frac{4}{11} = \frac{2}{11}$

3. $\frac{4}{5} + \frac{2}{5} = \frac{6}{5} = 1\frac{1}{5}$

4. $\frac{18}{25} - \frac{3}{25} = \frac{15}{25} = \frac{3}{5}$

5. $\frac{6}{11} + \frac{3}{11} = \frac{9}{11}$

6. $\frac{7}{8} - \frac{3}{8} = \frac{4}{8} = \frac{1}{2}$

7. $\frac{7}{15} + \frac{4}{15} = \frac{11}{15}$

8. $\frac{19}{24} - \frac{11}{24} = \frac{8}{24} = \frac{1}{3}$

9. $\frac{34}{45} + \frac{11}{45} = \frac{45}{45} = 1$

10. $\frac{17}{20} - \frac{9}{20} = \frac{8}{20} = \frac{2}{5}$

3·7

1. $\frac{3}{4} + \frac{2}{5} = \frac{15}{20} + \frac{8}{20} = \frac{23}{20} = 1\frac{3}{20}$

2. $\frac{7}{8} - \frac{1}{3} = \frac{21}{24} - \frac{8}{24} = \frac{13}{24}$

3. $\frac{5}{12} + \frac{3}{4} = \frac{5}{12} + \frac{9}{12} = \frac{14}{12} = 1\frac{2}{12} = 1\frac{1}{6}$

4. $\frac{14}{15} - \frac{7}{9} = \frac{42}{45} - \frac{35}{45} = \frac{7}{45}$

5. $\frac{3}{7} + \frac{5}{21} = \frac{9}{21} + \frac{5}{21} = \frac{14}{21} = \frac{2}{3}$

6. $\frac{17}{21} - \frac{2}{3} = \frac{17}{21} - \frac{14}{21} = \frac{3}{21} = \frac{1}{7}$

7. $\frac{2}{3} + \frac{1}{4} = \frac{8}{12} + \frac{3}{12} = \frac{11}{12}$

8. $\frac{6}{7} - \frac{5}{14} = \frac{12}{14} - \frac{5}{14} = \frac{7}{14} = \frac{1}{2}$

9. $\frac{5}{18} + \frac{1}{6} = \frac{5}{18} + \frac{3}{18} = \frac{8}{18} = \frac{4}{9}$

10. $\frac{3}{28} - \frac{3}{35} = \frac{15}{140} - \frac{12}{140} = \frac{3}{140}$

3·8

1. $\frac{64}{5} = 12\frac{4}{5}$

2. $\frac{19}{3} = 6\frac{1}{3}$

3. $\frac{46}{8} = 5\frac{6}{8} = 5\frac{3}{4}$

4. $\frac{140}{12} = 11\frac{8}{12} = 11\frac{2}{3}$

5. $\frac{14}{6} = 2\frac{2}{6} = 2\frac{1}{3}$

6. $1\frac{3}{8} = \frac{11}{8}$

7. $5\frac{1}{2} = \frac{11}{2}$

8. $3\frac{5}{7} = \frac{26}{7}$

9. $2\frac{3}{5} = \frac{13}{5}$

10. $9\frac{1}{3} = \frac{28}{3}$

3·9

1. $1\frac{1}{3} \times \frac{2}{5} = \frac{4}{3} \times \frac{2}{5} = \frac{8}{15}$

2. $3\frac{1}{2} \times \frac{2}{3} = \frac{7}{2} \times \frac{2}{3} = \frac{7}{3} = 2\frac{1}{3}$

3. $1\frac{3}{4} \times 1\frac{1}{3} = \frac{7}{4} \times \frac{4}{3} = \frac{7}{3} = 2\frac{1}{3}$

4. $5\frac{3}{8} \times 2\frac{2}{3} = \frac{43}{8} \times \frac{8}{3} = \frac{43}{3} = 14\frac{1}{3}$

5. $7\frac{1}{2} \times 3\frac{3}{5} = \frac{15}{2} \times \frac{18}{5} = \frac{27}{1} = 27$

6. $2\frac{1}{3} \div \frac{1}{6} = \frac{7}{3} \times \frac{6}{1} = \frac{14}{1} = 14$

7. $4\frac{1}{5} \div \frac{7}{10} = \frac{21}{5} \times \frac{10}{7} = \frac{6}{1} = 6$

8. $\frac{5}{7} \div 3\frac{3}{14} = \frac{5}{7} \div \frac{45}{14} = \frac{5}{7} \times \frac{14}{45} = \frac{2}{9}$

9. $6\frac{3}{7} \div 1\frac{4}{5} = \frac{45}{7} \div \frac{9}{5} = \frac{45}{7} \times \frac{5}{9} = \frac{25}{7} = 3\frac{4}{7}$

10. $12\frac{2}{3} \div 3\frac{1}{6} = \frac{38}{3} \div \frac{19}{6} = \frac{38}{3} \times \frac{6}{19} = \frac{4}{1} = 4$

3·10

1. $1\frac{3}{5} + 2\frac{1}{5} = 3\frac{4}{5}$

2. $5\frac{12}{13} - 3\frac{2}{13} = 2\frac{10}{13}$

3. $8\frac{5}{6} + 5\frac{5}{6} = 13\frac{10}{6} = 13 + 1\frac{4}{6} = 14\frac{4}{6} = 14\frac{2}{3}$

4. $9\frac{1}{12} - 5\frac{5}{12} = \left(8 + \frac{12}{12} + \frac{1}{12}\right) - 5\frac{5}{12} = 8\frac{13}{12} - 5\frac{5}{12} = 3\frac{8}{12} = 3\frac{2}{3}$

5. $5\frac{21}{32} + 4\frac{31}{32} = 9\frac{52}{32} = 9 + 1\frac{20}{32} = 10\frac{20}{32} = 10\frac{5}{8}$

6. $6\frac{3}{8} - 3\frac{5}{8} = \left(5 + \frac{8}{8} + \frac{3}{8}\right) - 3\frac{5}{8} = 5\frac{11}{8} - 3\frac{5}{8} = 2\frac{6}{8} = 2\frac{3}{4}$

7. $12\frac{3}{7} + 9\frac{1}{14} = 12\frac{6}{14} + 9\frac{1}{14} = 21\frac{7}{14} = 21\frac{1}{2}$

8. $5\frac{5}{6} - 1\frac{2}{3} = 5\frac{5}{6} - 1\frac{4}{6} = 4\frac{1}{6}$

9. $11\frac{3}{4} + 8\frac{5}{12} = 11\frac{9}{12} + 8\frac{5}{12} = 19\frac{14}{12} = 19 + 1\frac{2}{12} = 20\frac{1}{6}$

10. $8\frac{3}{11} - 4\frac{2}{5} = 8\frac{15}{55} - 4\frac{22}{55} = 7 + 1\frac{15}{55} - 4\frac{22}{55} = 7\frac{70}{55} - 4\frac{22}{55} = 3\frac{48}{55}$

11. $4\frac{9}{10} - 1\frac{2}{7} = 4\frac{63}{70} - 1\frac{20}{70} = 3\frac{43}{70}$

12. $5\frac{1}{6} + 3\frac{2}{5} = 5\frac{5}{30} + 3\frac{12}{30} = 8\frac{17}{30}$

13. $14\frac{9}{11} - 6\frac{2}{3} = 14\frac{27}{33} - 6\frac{22}{33} = 8\frac{5}{33}$

14. $\frac{19}{28} + 7\frac{3}{7} = \frac{19}{28} + 7\frac{12}{28} = 7\frac{31}{28} = 7 + 1\frac{3}{28} = 8\frac{3}{28}$

15. $11\frac{1}{3} - 4\frac{7}{8} = 11\frac{8}{24} - 4\frac{21}{24} = \left(10 + \frac{24}{24} + \frac{8}{24}\right) - 4\frac{21}{24} = 10\frac{32}{24} - 4\frac{21}{24} = 6\frac{11}{24}$

4 Decimals

1. Four and thirty-seven hundredths
2. Sixteen and twenty-five thousandths
3. Two and nine thousandths
4. Eight and seven thousand one hundred two ten-thousandths
5. One and four hundred twenty-six thousandths
6. 8.37
7. 4.0012
8. 5.00002
9. 300.045
10. 6.345

4·2

1. 0.75
2. 0.4
3. 0.64
4. 0.15
5. 0.1875

6. 0.07
7. 0.375
8. 0.21875
9. 0.0125
10. 0.055

4·3

1. $0.1\overline{3}$
2. $0.2\overline{7}$
3. $0.\overline{8}$
4. $0.\overline{6}$
5. $0.\overline{571428}$

6. $0.41\overline{6}$
7. $0.21\overline{6}$
8. $0.\overline{148}$
9. $0.0069\overline{4}$
10. $0.2\overline{3}$

4·4

1. $\frac{7}{10}$
2. $\frac{91}{100}$
3. $\frac{563}{1,000}$
4. $\frac{9}{20}$
5. $\frac{4}{25}$

6. $\frac{1}{200}$
7. $\frac{3}{250,000}$
8. $1\frac{17}{20}$
9. $2\frac{201}{5,000}$
10. $3\frac{253}{500}$

4·5

1. $\frac{1}{9}$
2. $\frac{2}{11}$
3. $\frac{1}{27}$
4. $\frac{5}{22}$
5. $\frac{2}{7}$

6. $\frac{3}{35}$
7. $\frac{1}{52}$
8. $\frac{2}{3}$
9. $\frac{6}{11}$
10. $\frac{7}{15}$

4·6

1. 47,000   48,000   48,000
2. 47,800   47,900   47,800
3. 40,000   50,000   50,000
4. 149.0   149.1   149.1
5. 149.0840   149.0841   149.0840
6. 9.436
7. 0.91

8. 15,000
9. 5,100
10. 410,000
11. 0.7338
12. 44,000
13. 339,000
14. 995.5
15. 2.00
16. $38,000 \times 20 = 760,000$
17. $3 \times 9 = 27$
18. $63 \div 9 = 7$
19. $600,000,000 \div 200,000 = 3,000$
20. $0.016 \div 0.004 = 4$

4·7

1. 7.961
2. 27.2901
3. 185.5
4. 63.0091
5. 949.2699
6. 19.196
7. 0.819
8. 698.7141
9. 11.4501
10. 30.9911

4·8

1. $3 \times 5 = 15$
2. $12 \times 30 = 360$
3. $80 \times 16 = 1,280$
4. $400 \times 3 = 1,200$
5. $400 \times 25 = 10,000$
6. 47.3
7. 2.744
8. 0.00324
9. 0.3024
10. 27.24908
11. 75.012
12. 13.082
13. 0.049995
14. 0.2500071
15. 7.2516318
16. 10.710
17. 1.0710
18. 10.710
19. 1.0710
20. 0.0010710

4·9

1. $24 \div 3 = 8$
2. $18 \div 12 = 1.5$
3. $72 \div 9 = 8$
4. $0.6 \div 3 = 0.2$
5. $66 \div 9 = 7\frac{1}{3}$
6. 388.75
7. 5.176
8. 2.59
9. $127.3\overline{1}$
10. $671.\overline{71}$
11. 2.5
12. 2.5
13. 0.025
14. 0.00025
15. 0.00000025

4·10

1. 940
2. 827,000
3. 163.2
4. 118,902
5. 41,487,000
6. 11.942
7. 1.85001
8. 0.00192
9. 0.0000603
10. 0.0121349

5 Ratios, proportions, and percents

5·1

1. Proportion $\dfrac{2}{3} = \dfrac{24}{36} \Rightarrow \dfrac{2}{3} = \dfrac{2 \times \cancel{12}}{3 \times \cancel{12}}$

2. Not a proportion $\dfrac{11}{15} = \dfrac{22}{33} \Rightarrow \dfrac{11}{15} = \dfrac{2 \times \cancel{11}}{3 \times \cancel{11}}$

3. Proportion $\dfrac{8}{15} = \dfrac{72}{135} \Rightarrow \dfrac{8}{15} = \dfrac{8 \times \cancel{9}}{15 \times \cancel{9}}$

4. Proportion $\dfrac{24}{75} = \dfrac{120}{375} \Rightarrow \dfrac{\cancel{3} \times 8}{\cancel{3} \times 25} = \dfrac{8 \times \cancel{15}}{25 \times \cancel{15}}$

5. Proportion $\dfrac{143}{169} = \dfrac{132}{156} \Rightarrow \dfrac{\cancel{13} \times 11}{\cancel{13} \times 13} = \dfrac{11 \times \cancel{12}}{13 \times \cancel{12}}$

6. Proportion $\dfrac{12}{21} = \dfrac{8}{14} \Rightarrow \underset{168}{\underline{12 \times 14}} = \underset{168}{\underline{21 \times 8}}$

7. Not a proportion $\dfrac{6}{8} = \dfrac{72}{98} \Rightarrow \underset{588}{\underline{6 \times 98}} \neq \underset{576}{\underline{8 \times 72}}$

8. Not a proportion $\dfrac{3}{11} = \dfrac{18}{65} \Rightarrow \underset{195}{\underline{3 \times 65}} \neq \underset{198}{\underline{11 \times 18}}$

9. Proportion $\dfrac{120}{150} = \dfrac{108}{135} \Rightarrow \underset{16,200}{\underline{120 \times 135}} = \underset{16,200}{\underline{150 \times 108}}$

10. Not a proportion $\dfrac{4}{123} = \dfrac{12}{41} \Rightarrow \underset{164}{\underline{4 \times 41}} \neq \underset{1476}{\underline{123 \times 12}}$

5·2
1. Proportion—swap
2. Proportion—flip
3. Not a proportion
4. Proportion—subtraction
5. Proportion—addition

6. Proportion—swap
7. Not a proportion
8. Proportion—equal ratios
9. Not a proportion
10. Not a proportion

5·3

1. $\dfrac{x}{15} = \dfrac{6}{9} \Rightarrow 9x = 90 \Rightarrow x = 10$

2. $\dfrac{12}{x} = \dfrac{96}{500} \Rightarrow 96x = 6{,}000 \Rightarrow x = 62.5$

3. $\dfrac{4}{7} = \dfrac{x}{49} \Rightarrow 7x = 196 \Rightarrow x = 28$

4. $\dfrac{3}{11} = \dfrac{13}{x} \Rightarrow 3x = 143 \Rightarrow 47\dfrac{2}{3}$

5. $\dfrac{9}{5} = \dfrac{x}{42} \Rightarrow 5x = 378 \Rightarrow x = 75.6$

6. $\dfrac{21}{x} = \dfrac{9}{42} \Rightarrow 9x = 882 \Rightarrow x = 98$

7. $\dfrac{x}{56} = \dfrac{3}{20} \Rightarrow 20x = 168 \Rightarrow x = 8.4$

8. $\dfrac{18}{20} = \dfrac{81}{x} \Rightarrow 18x = 1{,}620 \Rightarrow x = 90$

9. $\dfrac{x}{125} = \dfrac{40}{600} \Rightarrow 600x = 5{,}000 \Rightarrow x = 8\dfrac{1}{3}$

10. $\dfrac{91}{x} = \dfrac{21}{51} \Rightarrow 21x = 4{,}641 \Rightarrow x = 221$

5·4
1. $0.45 = 45\%$
2. $0.867 = 86.7\%$
3. $0.02 = 2\%$
4. $0.0003 = 0.03\%$
5. $0.15\overline{6} = 15.\overline{6}\%$

6. $0.\overline{49} = 49.\overline{49}\%$
7. $0.2\overline{3} = 0.2333... = 23.\overline{3}\%$
8. $4.583 = 458.3\%$
9. $2.9 = 290\%$
10. $0.\overline{123} = 0.123123... = 12.3123123....\% = 12.\overline{312}\%$

5·5
1. $29\% = 0.29$
2. $43.5\% = 0.435$
5. $54.\overline{54}\% = 0.\overline{54}$
6. $0.6\% = 0.006$
7. $129\% = 1.29$

3. $7\% = 0.07$
4. $33\dfrac{1}{3}\% = 33.\overline{3}\% = 0.\overline{3}$
8. $4.59\% = 0.0459$
9. $45.9\% = 0.459$
10. $459\% = 4.59$

5·6

1. $\dfrac{3}{4} = \dfrac{75}{100} = 75\%$

2. $\dfrac{1}{8} = 0.125 = 12.5\%$

3. $\dfrac{3}{5} = \dfrac{60}{100} = 60\%$

4. $\dfrac{2}{3} = 0.\overline{6} = 66.\overline{6}\% = 66\dfrac{2}{3}\%$

5. $\dfrac{4}{9} = 0.\overline{4} = 44.\overline{4}\% = 44\dfrac{4}{9}\%$

6. $\dfrac{3}{20} = \dfrac{15}{100} = 15\%$

7. $\dfrac{8}{15} = 0.5\overline{3} = 53.\overline{3}\% = 53\dfrac{1}{3}\%$

8. $\dfrac{7}{30} = 0.2\overline{3} = 23.\overline{3}\% = 23\dfrac{1}{3}\%$

9. $\dfrac{4}{45} = 0.08\overline{8} = 8.\overline{8}\% = 8\dfrac{8}{9}\%$

10. $\dfrac{5}{16} = 0.3125 = 31.25\%$

5·7

1. $15\% = \dfrac{15}{100} = \dfrac{3}{20}$

2. $120\% = \dfrac{120}{100} = \dfrac{6}{5} = 1\dfrac{1}{5}$

3. $2\% = \dfrac{2}{100} = \dfrac{1}{50}$

4. $0.75\% = 0.0075 = \dfrac{75}{10{,}000} = \dfrac{3}{400}$

5. $66.\overline{6}\% = 0.\overline{6} = \dfrac{6}{9} = \dfrac{2}{3}$

6. $4\dfrac{4}{9}\% = 4.\overline{4}\% = 0.0\overline{4} = \dfrac{4}{90} = \dfrac{2}{45}$

7. $0.1\% = 0.001 = \dfrac{1}{1{,}000}$

8. $0.\overline{1}\% = 0.00\overline{1} = \dfrac{1}{900}$

9. $1\% = \dfrac{1}{100}$

10. $100\% = \dfrac{100}{100} = 1$

5·8

1. $\dfrac{14}{70} = \dfrac{x}{100} \Rightarrow 70x = 1{,}400 \Rightarrow x = 20\%$

2. $\dfrac{x}{160} = \dfrac{15}{100} \Rightarrow 100x = 2{,}400 \Rightarrow x = 24$

182 Answer key

3. $\dfrac{27}{x} = \dfrac{80}{100} \Rightarrow 80x = 2{,}700 \Rightarrow x = 33.75$

4. $\dfrac{x}{1{,}492} = \dfrac{2}{100} \Rightarrow 100x = 2{,}984 \Rightarrow x = 29.84$

5. $\dfrac{254}{500} = \dfrac{x}{100} \Rightarrow 500x = 25{,}400 \Rightarrow 50.8\%$

6. $\dfrac{x}{849} = \dfrac{95}{100} \Rightarrow 100x = 80{,}655 \Rightarrow x = 806.55$

7. $\dfrac{x}{78} = \dfrac{250}{100} \Rightarrow 100x = 19{,}500 \Rightarrow x = 195$

8. $\dfrac{x}{200} = \dfrac{0.25}{100} \Rightarrow 100x = 50 \Rightarrow x = 0.5$

9. $\dfrac{\$30{,}000 - \$23{,}500}{\$30{,}000} = \dfrac{x}{100} \Rightarrow 30{,}000x = 650{,}000 \Rightarrow x = 21.\overline{6}\%$

10. $\dfrac{x}{\$50{,}000} = \dfrac{27}{100} \Rightarrow 100x = 1{,}350{,}000 \Rightarrow x = \$13{,}500$

11. $\dfrac{x}{38} = \dfrac{15}{100} \Rightarrow 100x = 570 \Rightarrow x = \$5.70$

12. $\dfrac{x}{\$75{,}000} = \dfrac{28}{100} \Rightarrow 100x = 2{,}100{,}000 \Rightarrow x = \$21{,}000$

13. $\dfrac{\$100 - \$80}{\$80} = \dfrac{x}{100} \Rightarrow 80x = 2{,}000 \Rightarrow x = 25\%$

14. Find the amount lost: $\dfrac{x}{\$20{,}000} = \dfrac{14}{100} \Rightarrow 100x = 280{,}000 \Rightarrow x = \$2{,}800$. Then subtract the loss from the original value. The car is worth $\$20{,}000 - \$2{,}800 = \$17{,}200$. Or, if the car loses 14% of its value, it keeps 86% of its value. $\dfrac{x}{\$20{,}000} = \dfrac{86}{100} \Rightarrow 100x = 1{,}720{,}000 \Rightarrow x = \$17{,}200$

15. Find the tax: $\dfrac{x}{120} = \dfrac{18}{100} \Rightarrow 100x = 2{,}160 \Rightarrow x = \$21.60$. Add that to the price of the room: $\$120 + \$21.60 = \$141.60$. Or realize that you'll pay 118% of the quoted rate: $\dfrac{x}{120} = \dfrac{118}{100} \Rightarrow 100x = 14{,}160 \Rightarrow x = \$141.60$

6   Exponents and roots

6·1

1. $6^2 = 6 \times 6 = 36$

2. $4^3 = 4 \times 4 \times 4 = 64$

3. $(-1)^{10} = 1$

4. $(-2)^2 = 4$

5. $\left(\dfrac{3}{4}\right)^2 = \dfrac{3}{4} \times \dfrac{3}{4} = \dfrac{9}{16}$

6. $-6^2 = -36$

7. $(-5)^3 = -125$

8. $2^{10} = 1{,}024$

9. $2.5^4 = 2.5 \times 2.5 \times 2.5 \times 2.5 = 39.0625$

10. $\left(\dfrac{1}{2}\right)^3 = \dfrac{1}{2} \times \dfrac{1}{2} \times \dfrac{1}{2} = \dfrac{1}{8}$

6·2

1. $2^3 \times 2^5 = 2^8$
2. $3^4 \times 3^7 = 3^{11}$
3. $9^5 \times 9^8 = 9^{13}$
4. $12^5 \times 12^3 = 12^6$

5. $x^2 \cdot x^3 = x^5$
6. $a^5 \times a^7 = a^{12}$
7. $2^5 \times 3^5 \times 3^8 = 2^5 \times 3^{13}$

8. $a^6 \times b^2 \times a^9 = a^{15} \times b^2$
9. $x^7 \cdot x^5 = x^{12}$
10. $z^{32} \cdot z^{15} = z^{47}$

6·3

1. $2^5 \div 2^3 = 2^2$

2. $3^7 \div 3^4 = 3^3$

3. $9^8 \div 9^5 = 9^3$

4. $12^5 \div 12^3 = 12^2$

5. $\dfrac{x^5}{x^2} = x^3$

6. $\dfrac{a^7}{a^5} = a^2$

7. $\dfrac{3^5 \times 3^8}{3^{11}} = \dfrac{3^{13}}{3^{11}} = 3^2$

8. $\dfrac{a^6 \times a^7}{a^9} = \dfrac{a^{13}}{a^9} = a^4$

9. $x^7 \div x^5 = x^2$

10. $z^{32} \div z^{15} = z^{17}$

6·4

1. $(4^3)^2 = 4^6$
2. $(5^2)^3 = 5^6$
3. $(7^3)^4 = 7^{12}$
4. $((-2)^3)^5 = (-2)^{15}$
5. $(21^2)^4 = 21^8$
6. $(x^3)^4 = x^{12}$
7. $(a^2)^9 = a^{18}$
8. $(y^8)^2 = y^{16}$
9. $(b^3)^8 = b^{24}$
10. $(v^4)^4 = v^{16}$

6·5

1. $(4 \times 2)^2 = 4^2 \times 2^2 = 16 \times 4 = 64$
2. $(-3 \times 10)^2 = (-3)^2 \times 10^2 = 9 \times 100 = 900$
3. $(2x^2)^4 = 2^4 x^8 = 16x^8$
4. $(3x^2)^3 = 3^3 x^6 = 27x^6$
5. $(-2a^3bc^4)^3 = (-2)^3 a^9 b^3 c^{12} = -8a^9 b^3 c^{12}$
6. $(2 \div 4)^3 = 2^3 \div 4^3 = 8 \div 64 = \dfrac{1}{8}$
7. $\left(\dfrac{7}{10}\right)^3 = \dfrac{7^3}{10^3} = \dfrac{343}{1,000}$
8. $\left(\dfrac{4x^5}{8x^3}\right)^2 = \dfrac{(4x^5)^2}{(8x^3)^2} = \dfrac{4^2 x^{10}}{8^2 x^6} = \dfrac{16x^{10}}{64x^6} = \dfrac{x^4}{4}$
9. $\left(\dfrac{3x^7}{5x^4}\right)^2 = \dfrac{(3x^7)^2}{(5x^4)^2} = \dfrac{3^2 x^{14}}{5^2 x^8} = \dfrac{9x^{14}}{25x^8} = \dfrac{9x^6}{25}$
10. $\left(\dfrac{6a^2bc^4}{3abc}\right)^5 = \dfrac{6^5 a^{10} b^5 c^{20}}{3^5 a^5 b^5 c^5} = \dfrac{7,776 a^{10} b^5 c^{20}}{243 a^5 b^5 c^5} = 32a^5 c^{15}$

6·6

1. $(2 \times 4)^{-1} = \dfrac{1}{8}$
2. $\left(\dfrac{2}{3}\right)^{-3} = \left(\dfrac{3}{2}\right)^3 = \dfrac{27}{8}$
3. $y^{-2} y^3 y^{-4} = y^{-3} = \dfrac{1}{y^3}$
4. $(x^4 y^{-4})^0 = (2x^{-2}) = 1 \cdot 2x^{-2} = \dfrac{2}{x^2}$
5. $\dfrac{12x^5 y^8}{4x^{-2} y^{-4}} = 3x^7 y^{12}$
6. $(-2x^{-1} y^{-3})^3 = (-2)^3 x^{-3} y^{-9} = \dfrac{-8}{x^3 y^9}$
7. $(4x^3 y^{-45})^4 (4x^3 y^{-45})^{-4} = 4^4 x^{12} y^{-180} \cdot 4^{-4} x^{-12} y^{180} = 4^0 x^0 y^0 = 1$
8. $\left(\dfrac{2x^3}{z^4}\right)^{-3} = \left(\dfrac{z^4}{2x^3}\right)^3 = \dfrac{z^{12}}{2^3 x^9} = \dfrac{z^{12}}{8x^9}$
9. $\dfrac{15x^{-5} y^2}{2x^5 y} \cdot \dfrac{xy^{-2}}{5x^3 y^{-2}} = \dfrac{\overset{3}{\cancel{15}} x^{-10} y}{2} \cdot \dfrac{1}{\cancel{5} x^2} = \dfrac{3}{2} x^{-12} y = \dfrac{3y}{2x^{12}}$
10. $\left(\dfrac{3x^2 y}{2x^{-3} y^2}\right)^{-2} = \left(\dfrac{2x^{-3} y^2}{3x^2 y}\right)^2 = \dfrac{4x^{-6} y^4}{9x^4 y^2} = \dfrac{4y^2}{9x^{10}}$

6·7

1. $83,681 + 119,392 \approx 80,000 + 120,000 = 200,000$
2. $9,384,583 - 5,639,299 \approx 9,000,000 - 6,000,000 = 3,000,000$
3. $62,973 \div 8,856 \approx 63,000 \div 9,000 = 7$
4. $293,449 \times 781,772 \approx 300,000 \times 800,000 = 240,000,000,000$
5. $11,735,882 \times 49,948,221 \approx 12,000,000 \times 50,000,000 = 600,000,000,000,000$
6. $9.29 \times 10^6 = 9,290,000$
7. $1.35 \times 10^{-3} = 0.00135$
8. $4.73 \times 10^{12} = 4,730,000,000,000$
9. $0.000234 = 2.34 \times 10^{-4}$
10. $87,000 = 8.7 \times 10^4$
11. $972,734,000,000 = 9.72734 \times 10^{11}$
12. $(5 \times 10^{-3})(3 \times 10^6) = 15 \times 10^3 = 1.5 \times 10^4$
13. $\dfrac{7.7 \times 10^{-4}}{1.1 \times 10^{-1}} = 7 \times 10^{-3}$
14. $(5 \times 10^{-2})^3 = 5^3 \times 10^{(-2 \times 3)} = 125 \times 10^{-6} = 1.25 \times 10^{-4}$
15. $(4.2 \times 10^{-8})(3 \times 10^5) = 12.6 \times 10^{-3} = 1.26 \times 10^{-2}$

6·8

1. $\sqrt{128} = \sqrt{64 \times 2} = 8\sqrt{2}$

2. $\sqrt{675} = \sqrt{25 \times 27} = \sqrt{25 \times 9 \times 3} = 5 \times 3\sqrt{3} = 15\sqrt{3}$

3. $\sqrt{180} = \sqrt{36 \times 5} = 6\sqrt{5}$

4. $\sqrt{1,225} = \sqrt{25 \times 49} = 5 \times 7 = 35$

5. $\dfrac{1}{\sqrt{2}} \times \dfrac{\sqrt{2}}{\sqrt{2}} = \dfrac{\sqrt{2}}{2}$

6. $\dfrac{12}{\sqrt{6}} \times \dfrac{\sqrt{6}}{\sqrt{6}} = \dfrac{12\sqrt{6}}{6} = 2\sqrt{6}$

7. $\dfrac{2\sqrt{5}}{5\sqrt{2}} \times \dfrac{\sqrt{2}}{\sqrt{2}} = \dfrac{2\sqrt{10}}{5 \times 2} = \dfrac{\sqrt{10}}{5}$

8. $\sqrt{75x^2 y^4} = \sqrt{25 \cdot 3 \cdot x^2 \cdot y^2 \cdot y^2} = 5xy^2\sqrt{3}$

9. $\sqrt{80a^3 b^4 c^2} = \sqrt{16 \cdot 5 \cdot a^2 \cdot a \cdot b^2 \cdot b^2 \cdot c^2} = 4ab^2 c\sqrt{5a}$

10. $\sqrt{\dfrac{120x^2 y}{15xy^3}} = \sqrt{\dfrac{8x}{y^2}} = \dfrac{2\sqrt{2x}}{y}$

6·9

1. $\sqrt{12} \cdot \sqrt{27} = 2\sqrt{3} \cdot 3\sqrt{3} = 6 \cdot 3 = 18$

2. $\sqrt{5} \cdot \sqrt{35} = \sqrt{5 \cdot 5 \cdot 7} = 5\sqrt{7}$

3. $2\sqrt{3} \cdot \sqrt{6} = 2\sqrt{18} = 2 \times 3\sqrt{2} = 6\sqrt{2}$

4. $4\sqrt{30} \cdot 5\sqrt{33} = 20\sqrt{990} = 20\sqrt{9 \times 110} = 20 \times 3\sqrt{110} = 60\sqrt{110}$

5. $-5\sqrt{45} \cdot 8\sqrt{75} = -40\sqrt{9 \times 5 \times 3 \times 25} = -40 \times 3 \times 5\sqrt{5 \times 3} = -600\sqrt{15}$

6. $\dfrac{18\sqrt{15}}{6\sqrt{5}} = 3\sqrt{3}$

7. $\dfrac{9\sqrt{162}}{27\sqrt{8}} = \dfrac{\sqrt{162}}{3\sqrt{8}} = \dfrac{\sqrt{81 \times 2}}{3\sqrt{4 \times 2}} = \dfrac{9\sqrt{2}}{6\sqrt{2}} = \dfrac{3}{2}$

8. $\dfrac{32\sqrt{75}}{8\sqrt{15}} = 4\sqrt{5}$

9. $\dfrac{10\sqrt{6}}{\sqrt{8}} = \dfrac{10\sqrt{6}}{2\sqrt{2}} = 5\sqrt{3}$

10. $\dfrac{7\sqrt{288}}{21\sqrt{48}} = \dfrac{\sqrt{288}}{3\sqrt{48}} = \dfrac{\sqrt{144 \times 2}}{3\sqrt{16 \times 3}} = \dfrac{12\sqrt{2}}{12\sqrt{3}} = \dfrac{\sqrt{2}}{\sqrt{3}} \times \dfrac{\sqrt{3}}{\sqrt{3}} = \dfrac{\sqrt{6}}{3}$

6·10

1. $\sqrt{8} + \sqrt{72} = 2\sqrt{2} + 6\sqrt{2} = 8\sqrt{2}$

2. $\sqrt{12} + 2\sqrt{48} = 2\sqrt{3} + 2 \cdot 4\sqrt{3} = 2\sqrt{3} + 8\sqrt{3} = 10\sqrt{3}$

3. $2\sqrt{75} + 9\sqrt{147} = 2 \cdot 5\sqrt{3} + 9 \cdot 7\sqrt{3} = 10\sqrt{3} + 63\sqrt{3} = 73\sqrt{3}$

4. $\sqrt{18} - 3\sqrt{50} + 5\sqrt{8} = 3\sqrt{2} - 3 \cdot 5\sqrt{2} + 5 \cdot 2\sqrt{2} = 3\sqrt{2} - 15\sqrt{2} + 10\sqrt{2} = -2\sqrt{2}$

5. $12\sqrt{405} + 7\sqrt{500} - 3\sqrt{180} = 12 \cdot 9\sqrt{5} + 7 \cdot 10\sqrt{5} - 3 \cdot 6\sqrt{5}$
$$= 108\sqrt{5} + 70\sqrt{5} - 18\sqrt{5} = 160\sqrt{5}$$

6. $\dfrac{12}{\sqrt{6}} + \sqrt{6} = \dfrac{12}{\sqrt{6}} \cdot \dfrac{\sqrt{6}}{\sqrt{6}} + \sqrt{6} = \dfrac{12\sqrt{6}}{6} + \sqrt{6} = 2\sqrt{6} + \sqrt{6} = 3\sqrt{6}$

7. $\dfrac{4}{2\sqrt{2}} + \dfrac{7\sqrt{2}}{4} = \dfrac{4}{2\sqrt{2}} \cdot \dfrac{\sqrt{2}}{\sqrt{2}} + \dfrac{7\sqrt{2}}{4} = \dfrac{4\sqrt{2}}{4} + \dfrac{7\sqrt{2}}{4} = \dfrac{11\sqrt{2}}{4}$

8. $\dfrac{8\sqrt{2}}{6\sqrt{3}} + \dfrac{7\sqrt{3}}{9\sqrt{2}} = \dfrac{8\sqrt{2}}{6\sqrt{3}} \cdot \dfrac{\sqrt{3}}{\sqrt{3}} + \dfrac{7\sqrt{3}}{9\sqrt{2}} \cdot \dfrac{\sqrt{2}}{\sqrt{2}} = \dfrac{8\sqrt{6}}{18} + \dfrac{7\sqrt{6}}{18} = \dfrac{15\sqrt{6}}{18} = \dfrac{5\sqrt{6}}{6}$

9. $\dfrac{4}{3\sqrt{5}} - \dfrac{2}{\sqrt{45}} = \dfrac{4}{3\sqrt{5}} - \dfrac{2}{3\sqrt{5}} = \dfrac{2}{3\sqrt{5}} \cdot \dfrac{\sqrt{5}}{\sqrt{5}} = \dfrac{2\sqrt{5}}{15}$

10. $\dfrac{6\sqrt{7}}{2\sqrt{5}} - \dfrac{7\sqrt{5}}{4\sqrt{7}} = \dfrac{6\sqrt{7}}{2\sqrt{5}} \cdot \dfrac{4\sqrt{7}}{4\sqrt{7}} - \dfrac{7\sqrt{5}}{4\sqrt{7}} \cdot \dfrac{2\sqrt{5}}{2\sqrt{5}} = \dfrac{24 \cdot 7 - 14 \cdot 5}{8\sqrt{35}} = \dfrac{98}{8\sqrt{35}} \cdot \dfrac{\sqrt{35}}{\sqrt{35}} = \dfrac{98\sqrt{35}}{8 \cdot 35} = \dfrac{7\sqrt{35}}{20}$

7 Equations and inequalities

7·1

1. $x + 7 = 12 \Rightarrow x = 5$

2. $y - 5 = 16 \Rightarrow y = 21$

3. $t + 5 = 16 \Rightarrow t = 11$

4. $w - 13 = 34 \Rightarrow w = 47$

5. $x + \dfrac{1}{2} = \dfrac{7}{2} \Rightarrow x = \dfrac{6}{2} = 3$

6. $z - 2.9 = 3.1 \Rightarrow z = 6.0$

7. $y - 4\dfrac{3}{4} = 7\dfrac{1}{2} \Rightarrow y = 12\dfrac{1}{4}$

8. $x + 10 = 6 \Rightarrow x = -4$

9. $y - 12 = -4 \Rightarrow y = 8$

10. $t + 11 = -4 \Rightarrow t = -15$

11. $8x = 48 \Rightarrow x = 6$

12. $\dfrac{z}{7} = 8 \Rightarrow z = 56$

13. $-6y = 42 \Rightarrow y = -7$

14. $\dfrac{t}{9} = -7 \Rightarrow t = -63$

15. $15x = 45 \Rightarrow x = 3$

16. $\dfrac{w}{11} = 23 \Rightarrow w = 253$

17. $\dfrac{4}{5}t = \dfrac{8}{15} \Rightarrow t = \dfrac{2}{3}$

18. $\dfrac{m}{4.3} = -3.1 \Rightarrow m = -13.33$

19. $-1.4x = 4.2 \Rightarrow x = -3$

20. $\dfrac{z}{7} = -35 \Rightarrow z = -245$

7·2

1. $3x - 11 = 37 \Rightarrow x = 16$

2. $-8t + 7 = 23 \Rightarrow t = -2$

3. $4 - 3x = -11 \Rightarrow x = 5$

4. $5 + 3x = 26 \Rightarrow x = 7$

5. $\dfrac{x}{4} - 8 = -5 \Rightarrow x = 12$

6. $-5x + 19 = -16 \Rightarrow x = 7$

7. $\dfrac{x}{3} + 7 = 16 \Rightarrow x = 27$

8. $13 - 4x = -15 \Rightarrow x = 7$

9. $\dfrac{x}{2} - \dfrac{3}{4} = \dfrac{1}{4} \Rightarrow x = 2$

10. $12x - 5 = -53 \Rightarrow x = -4$

7·3

1. $3t + 8t = 11t$
2. $10x - 6x = 4x$
3. $5x + 3y - 2x = 3x + 3y$
4. $2y - 3 + 5x + 8y - 4x = x + 10y - 3$
5. $6 - 3x + x^2 - 7 + 5x - 3x^2 = -1 + 2x - 2x^2$

6. $(5t + 3) + (t - 12r) - 8 + 9r + (7t - 5) = 5t + 3 + t - 12r - 8 + 9r + 7t - 5$
$$= 13t - 3r - 10$$

7. $(5x^2 - 9x + 7) + (2x^2 + 3x + 12) = 5x^2 - 9x + 7 + 2x^2 + 3x + 12$
$$= 7x^2 - 6x + 19$$

8. $(2x - 7) - (y + 2x) - (3 + 5y) + (8x - 9) = 2x - 7 - y - 2x - 3 - 5y + 8x - 9$
$$= 8x - 6y - 19$$

9. $(3x^2 + 5x - 3) - (x^2 + 3x - 4) = 3x^2 + 5x - 3 - x^2 - 3x + 4$
$$= 2x^2 + 2x + 1$$

10. $2y - (3 + 5x) + 8y - (4x - 3) = 2y - 3 - 5x + 8y - 4x + 3$
$$= 10y - 9x$$

7·4

1. $5x - 8 = x + 22 \Rightarrow 4x - 8 = 22 \Rightarrow 4x = 30 \Rightarrow x = 7.5$
2. $11x + 18 = 7x - 14 \Rightarrow 4x + 18 = -14 \Rightarrow x = -8$
3. $3x + 18 = 4x - 9 \Rightarrow -x + 18 = -9 \Rightarrow x = 27$
4. $30 - 4x = 16 + 3x \Rightarrow 30 - 7x = 16 \Rightarrow x = 2$
5. $11x - 5 = 10 - 4x \Rightarrow 15x - 5 = 10 \Rightarrow x = 1$
6. $8x - 13 = 12 + 3x \Rightarrow 5x - 13 = 12 \Rightarrow x = 5$
7. $3x - 5 = 2 - 4x \Rightarrow 7x - 5 = 2 \Rightarrow x = 1$
8. $1.5x - 7.1 = 8.4 + x \Rightarrow 0.5x - 7.1 = 8.4 \Rightarrow x = 31$
9. $13 - 9x = 7x - 19 \Rightarrow 13 - 16x = -19 \Rightarrow x = 2$
10. $-5x + 21 = 27 - x \Rightarrow -4x + 21 = 27 \Rightarrow x = -1.5$

7·5

1. $5(x + 2) = 40 \Rightarrow 5x + 10 = 40 \Rightarrow x = 6$
2. $4(x - 7) + 6 = 18 \Rightarrow 4x - 22 = 18 \Rightarrow x = 10$
3. $5(x - 4) = 7(x - 6) \Rightarrow 5x - 20 = 7x - 42 \Rightarrow x = 11$
4. $4(5x + 3) + x = 6(x + 2) \Rightarrow 21x + 12 = 12x + 12 \Rightarrow x = 0$
5. $8(x - 4) - 16 = 10(x - 7) \Rightarrow 8x - 48 = 10x - 70 \Rightarrow x = 11$

6. $6(2x+9)-30=4(7x-2) \Rightarrow 12x+24=28x-8 \Rightarrow x=2$

7. $7(x-1)+2x=12+5(x+1) \Rightarrow 9x-7=5x+17 \Rightarrow x=6$

8. $6(x-1)-2x=2(x+1)+4(2-x) \Rightarrow 4x-6=-2x+10 \Rightarrow x=2\frac{2}{3}$

9. $5(6x+2)+7(4-12x)=35-(6+27x) \Rightarrow -54x+38=29-27x \Rightarrow x=\frac{1}{3}$

10. $8(2x-5)-2(x-2)=5(x+7)-4(x+8) \Rightarrow 14x-36=x+3 \Rightarrow x=3$

7·6

1. $3x-5 \geq 22 \Rightarrow x \geq 9$

2. $2x-5 > 13-4x \Rightarrow x > 3$

3. $3x+2 \leq 8x+22 \Rightarrow x \geq -4$

4. $12x+3 < x+36 \Rightarrow x < 3$

5. $t-9 \geq 24-10t \Rightarrow t \geq 3$

6. $2y-13 > 4(2-y) \Rightarrow y > 3.5$

7. $5x+10(x-1) \geq 95 \Rightarrow x \geq 7$

8. $5x-4 \leq 13x+28 \Rightarrow x \geq -4$

9. $3x-2 < 2x-3 \Rightarrow x < -1$

10. $-x+5 \geq -2+x \Rightarrow x \leq 3.5$

8   Quadratic equations

8·1

1. $x^2=64 \Rightarrow x=\pm 8$

2. $x^2-16=0 \Rightarrow x=\pm 4$

3. $x^2-8=17 \Rightarrow x=\pm 5$

4. $x^2=18 \Rightarrow x=\pm 3\sqrt{2}$

5. $3x^2=48 \Rightarrow x=\pm 4$

6. $t^2-1{,}000=0 \Rightarrow t=\pm 10\sqrt{10}$

7. $2y^2-150=0 \Rightarrow y=\pm 5\sqrt{3}$

8. $9x^2=4 \Rightarrow x=\pm\frac{2}{3}$

9. $64y^2=25 \Rightarrow y=\pm\frac{5}{8}$

10. $4x^2-15=93 \Rightarrow x=\pm 3\sqrt{3}$

8·2

1. $(x+8)(x+2)=x^2+10x+16$

2. $(y-4)(y-9)=y^2-13y+36$

3. $(t-2)(t+6)=t^2+4t-12$

4. $(2x+8)(x-3)=2x^2+2x-24$

5. $(y-9)(3y+1)=3y^2-26y-9$

6. $(5x-6)(3x+4)=15x^2+2x-24$

7. $(6x-1)(x+5)=6x^2+29x-5$

8. $(1-3b)(5+2b)=5-13b-6b^2$

9. $(3x-7)(2x+5)=6x^2+x-35$

10. $(5-2x)(5x-2)=-10x^2+29x-10$

8·3

1. $x^2+12x+35=(x+7)(x+5)$

2. $x^2+11x+28=(x+7)(x+4)$

3. $x^2-8x+15=(x-5)(x-3)$

4. $x^2-7x+12=(x-4)(x-3)$

5. $x^2+x-20=(x+5)(x-4)$

6. $x^2-2x-3=(x-3)(x+1)$

7. $x^2-11x+18=(x-9)(x-2)$

8. $x^2-9x-22=(x-11)(x+2)$

9. $x^2+10x-39=(x+13)(x-3)$

10. $x^2+12x+32=(x+8)(x+4)$

8·4

1. $x^2+5x+6=0 \Rightarrow (x+3)(x+2)=0 \Rightarrow x=-3, x=-2$

2. $x^2+12=7x \Rightarrow x^2-7x+12=0 \Rightarrow (x-4)(x-3)=0 \Rightarrow x=4, x=3$

3. $y^2+3y=8+y \Rightarrow y^2+2y-8=0 \Rightarrow (y+4)(y-2)=0 \Rightarrow y=-4, y=2$

4. $a^2-3a-4=6 \Rightarrow a^2-3a-10=0 \Rightarrow (a-5)(a+2)=0 \Rightarrow a=5, a=-2$

5. $20=x^2+x \Rightarrow x^2+x-20=0 \Rightarrow (x+5)(x-4)=0 \Rightarrow x=-5, x=4$

6. $x^2+5=6x \Rightarrow x^2-6x+5=0 \Rightarrow (x-5)(x-1)=0 \Rightarrow x=5, x=1$

7. $x^2+3x=0 \Rightarrow x(x+3)=0 \Rightarrow x=0, x=-3$

8. $x^2=5x \Rightarrow x^2-5x=0 \Rightarrow x(x-5)=0 \Rightarrow x=0, x=5$

9. $2x^2-1=3x^2-2x \Rightarrow x^2-2x+1=0 \Rightarrow (x-1)(x-1)=0 \Rightarrow x=1, x=1$

10. $2x^2-x=0 \Rightarrow x(2x-1)=0 \Rightarrow x=0, x=\frac{1}{2}$

8·5

1. $x^2+4x-21=0 \Rightarrow x=\dfrac{-4\pm\sqrt{4^2-4\cdot 1\cdot(-21)}}{2\cdot 1} \Rightarrow x=3, x=-7$

2. $t^2=10-3t \Rightarrow t=\dfrac{-3\pm\sqrt{3^2-4\cdot 1\cdot(-10)}}{2\cdot 1} \Rightarrow t=2, t=-5$

3. $y^2-4y=32 \Rightarrow y=\dfrac{-(-4)\pm\sqrt{(-4)^2-4\cdot 1\cdot(-32)}}{2\cdot 1} \Rightarrow y=8, y=-4$

4. $x^2=6+x \Rightarrow x=\dfrac{-(-1)\pm\sqrt{(-1)^2-4\cdot 1\cdot(-6)}}{2\cdot 1} \Rightarrow x=3, x=-2$

5. $6x+x^2=9 \Rightarrow x=\dfrac{-6\pm\sqrt{6^2-4\cdot 1\cdot(-9)}}{2\cdot 1} \Rightarrow x=\dfrac{-6\pm 6\sqrt{2}}{2}=-3\pm 3\sqrt{2}$

6. $t^2 + 6t - 15 = 0 \Rightarrow t = \dfrac{-6 \pm \sqrt{6^2 - 4 \cdot 1 \cdot (-15)}}{2 \cdot 1} \Rightarrow t = \dfrac{-6 \pm 4\sqrt{6}}{2} = -3 \pm 2\sqrt{6}$

7. $4x^2 - 3 = x \Rightarrow x = \dfrac{-(-1) \pm \sqrt{(-1)^2 - 4 \cdot 4 \cdot (-3)}}{2 \cdot 4} \Rightarrow x = 1, x = -\dfrac{3}{4}$

8. $3x^2 - 1 = 2x \Rightarrow x = \dfrac{-(-2) \pm \sqrt{(-2)^2 - 4 \cdot 3 \cdot (-1)}}{2 \cdot 3} \Rightarrow x = 1, x = -\dfrac{1}{3}$

9. $x + 5 = 3x^2 - x \Rightarrow x = \dfrac{-(-2) \pm \sqrt{(-2)^2 - 4 \cdot 3 \cdot (-5)}}{2 \cdot 3} \Rightarrow x = 1\dfrac{2}{3}, x = -1$

10. $6x^2 - 2 = x \Rightarrow x = \dfrac{-(-1) \pm \sqrt{(-1)^2 - 4 \cdot 6 \cdot (-2)}}{2 \cdot 6} \Rightarrow x = \dfrac{2}{3}, x = -\dfrac{1}{2}$

9 Segments and angles

9·1

1. $LP = 12$
2. $MO = 6$
3. $NQ = 10$
4. $NM = 2$
5. $QL = 15$
6. $MP = 9$
7. $ON = 4$
8. $\overline{LM} \cong \overline{OP}$ or PQ
9. $\overline{MO} \cong \overline{OQ}$
10. $\overline{MP} \cong \overline{LO}$
11. The midpoint of $\overline{OQ}$ is P.
12. If A is the midpoint of $\overline{NO}$, A is -2.
13. If B is the midpoint of $\overline{MN}$, B is -5.
14. If M is the midpoint of $\overline{LC}$, C is -3.
15. If P is the midpoint of $\overline{ND}$, D is 10.

9·2

1. $\angle 1 = \angle QRV$ or $\angle PRW$
2. $\angle 4 = \angle WVS$
3. $\angle QRV$, $\angle VRS$, and $\angle QRS$
4. $\angle 6 = \angle XQP$
5. $\angle TSZ = \angle 3$
6. The vertex of $\angle 5$ is point Y.
7. The sides of $\angle 2$ are $\overrightarrow{RV}$ and $\overrightarrow{RS}$.
8. True: $\angle WVS$ is another name for $\angle WVZ$. S and Z are both on the same line.
9. False: $\angle YQV$ is not another name for $\angle 5$. The vertex of $\angle 5$, Y, must be the middle letter.
10. False: $\angle RVY$ is not another name for $\angle 4$. $\angle 4$ has its vertex at V but its sides don't go through R and Y.
11. $\angle AOF = 120°$
12. $\angle BOE = 75°$
13. $\angle COD = 30°$
14. $\angle DOB = 50°$
15. $\angle EOG = 60°$
16. $\angle FOB = 100°$
17. $\angle GOC = 115°$
18. $\angle EOA = 95°$
19. $\angle COB = 20°$
20. $\angle DOG = 85°$
21. False: $m\angle TPU = m\angle WPX$, but $\overrightarrow{PU}$ is NOT the bisector of $\angle TPX$. $\angle TPU$ and $\angle WPX$ are angles of equal size, but they don't make up two halves of $\angle TPX$. $\angle TPX$ includes more than those two angles.
22. False: $\overrightarrow{PT}$ is NOT the bisector of $\angle SPU$ because $m\angle SPT \neq m\angle TPU$
23. The bisector of $\angle QPS$ is PR.
24. The coordinate of the bisector of $\angle TPW$ is 100.
25. The coordinate of the bisector of $\angle RPU$ is 50.

9·3

1. $\angle RPS$ is acute.
2. $\angle QPY$ is straight.
3. $\angle UPW$ is acute.
4. $\angle TPW$ is right.
5. $\angle QPR$ is acute.
6. $\angle UPX$ is acute.
7. $\angle RPV$ is obtuse.
8. $\angle QPX$ is obtuse.

9. $\angle WPX$ is acute.

10. $\angle XPR$ is obtuse.

11. The measure of the complement of an angle of 53° is 37°.

12. The measure of the complement of an angle of 31° is 59°.

13. The measure of the supplement of an angle of 47° is 133°.

14. The measure of the supplement of an angle of 101° is 79°.

15. $\angle HCI$ and $\angle ACG$ are vertical angles.

16. $\angle EAD$ and $\angle DAG$ are a linear pair.

17. $\angle GBA$ and $\angle LBK$ are vertical angles.

18. $\angle GBK$ and $\angle LBK$ are a linear pair.

19. $\angle BAG$ is congruent to $\angle DAE$.

20. $\angle JGC$ is supplementary to $\angle CGA$.

10   Coordinate geometry

10·1

1. $d = 5$
2. $d = 15$
3. $d = 13$
4. $d = 3\sqrt{2} \approx 4.24$

5. $d = 7$
6. $(4, 6)$
7. $(-2, 4)$

8. $(-3, -2)$
9. $(4, 4)$
10. $(2, -3)$

10·2

1. $m = -\dfrac{3}{5}$

2. $m = -\dfrac{2}{3}$

3. $m = 0$

4. $m = \dfrac{1}{4}$

5. $m = -3$

6. $y = 9x + 4$ Slope: 9, y-intercept: 4

7. $y = -3x + 2$ Slope: -3, y-intercept: 2

8. $y + 5 = \dfrac{1}{2}x \Rightarrow y = \dfrac{1}{2}x - 5$ Slope: $\dfrac{1}{2}$, y-intercept: -5

9. $x + y = 6 \Rightarrow y = -x + 6$ Slope: -1, y-intercept: 6

10. $2x - 3y = 12 \Rightarrow y = \dfrac{2}{3}x - 4$ Slope: $\dfrac{2}{3}$, y-intercept: -4

11. $y = -2x + 7$

12. $y = \dfrac{1}{2}x - 9$

13. $y = -\dfrac{1}{5}x + \dfrac{2}{3}$

14. $y + 1 = 5(x - 3) \Rightarrow y = 5x - 16$

15. $y - 1 = -3(x - 1) \Rightarrow y = -3x + 4$

16. $y - 13 = 4(x - 4) \Rightarrow y = 4x - 3$

17. $y - 0 = 2(x - 8) \Rightarrow y = 2x - 16$

18. $y - 5 = \dfrac{3}{4}(x + 4) \Rightarrow y = \dfrac{3}{4}x + 8$

19. $m = \dfrac{3 + 5}{6 - 4} = 4 \Rightarrow y - 3 = 4(x - 6) \Rightarrow y = 4x - 21$

20. $m = \dfrac{11 + 1}{5 + 1} = 2 \Rightarrow y - 11 = 2(x - 5) \Rightarrow y = 2x + 1$

1. $y = -\dfrac{3}{4}x + 1$

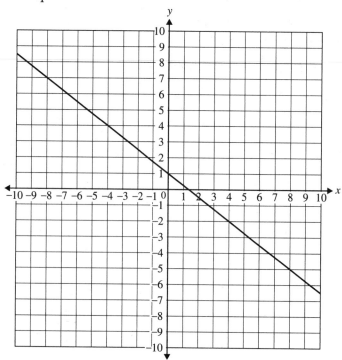

2. $2x - 3y = 9 \Rightarrow y = \dfrac{2}{3}x - 3$

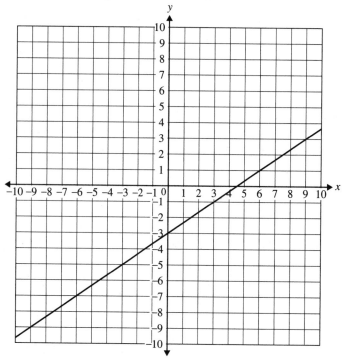

3. $y = -4x + 6$

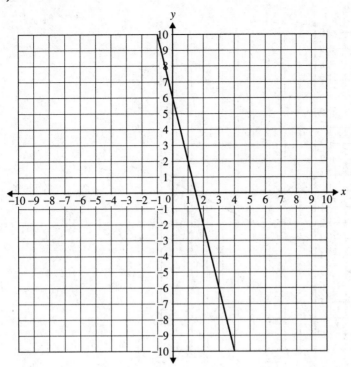

4. $6x + 2y = 18 \Rightarrow y = -3x + 9$

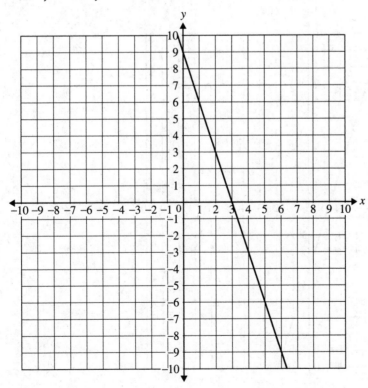

5. $x - 2y = 8 \Rightarrow y = \dfrac{1}{2}x - 4$

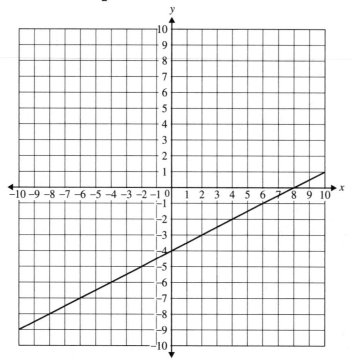

6. $y = -3x - 4$

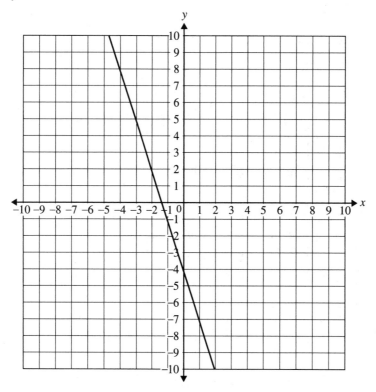

7. $y - 6 = 3x + 1 \Rightarrow y = 3x + 7$

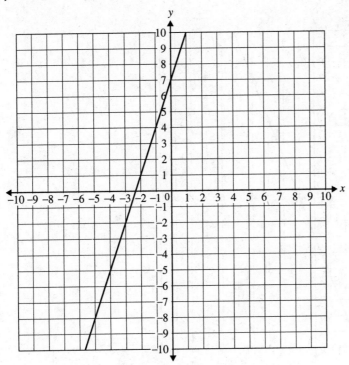

8. $3x + 5y = 15 \Rightarrow y = -\dfrac{3}{5}x + 3$

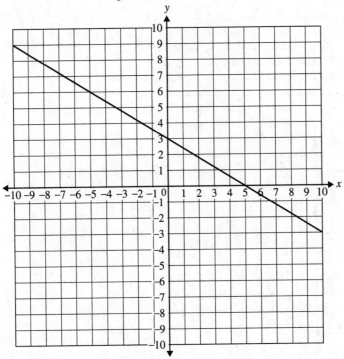

9. $2y = 5x - 6 \Rightarrow y = \dfrac{5}{2}x - 3$

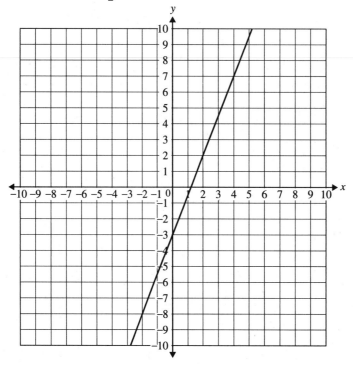

10. $3x - 2y - 6 = 0 \Rightarrow y = \dfrac{3}{2}x - 3$

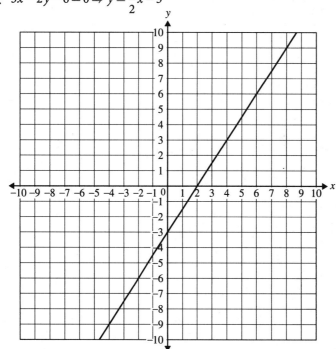

10·4

1. $y = x^2 - 4x + 3 = (x-3)(x-1)$ x-intercepts: $(3, 0)$, $(1, 0)$, y-intercept: $(0, 3)$
2. $y = x^2 - 4x - 5 = (x-5)(x+1)$ x-intercepts: $(5, 0)$, $(-1, 0)$, y-intercept: $(0, -5)$
3. $y = x^2 + 2x = x(x+2)$ x-intercepts: $(0, 0)$, $(-2, 0)$, y-intercept: $(0, 0)$
4. $y = x^2 - 7x + 12 = (x-4)(x-3)$ x-intercepts: $(4, 0)$, $(3, 0)$, y-intercept: $(0, 12)$
5. $y = x^2 - 2x + 1 = (x-1)(x-1)$ x-intercepts: $(1, 0)$, y-intercept: $(0, 1)$
6. $y = x^2 - 8x + 15$ Vertex: $(4, -1)$
7. $y = x^2 + 4x - 2$ Vertex: $(-2, -6)$
8. $y = 2x^2 - 4x + 3$ Vertex: $(1, 1)$
9. $y = -x^2 + 6x - 7$ Vertex: $(3, 2)$
10. $y = -x^2 + 4x + 7$ Vertex: $(2, 11)$
11. $y = 2x^2 - 1$

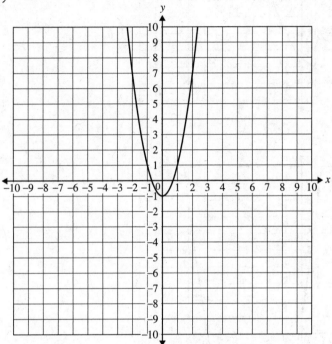

12. $y = -x^2 + 8x$

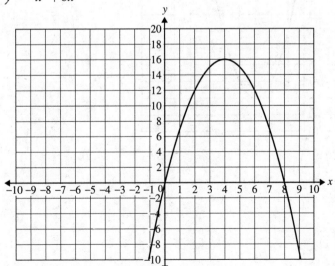

13. $y = x^2 + 2x - 15$

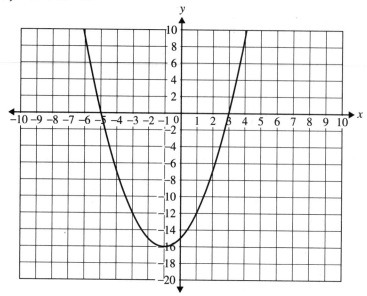

14. $y = x^2 - 6x + 1$

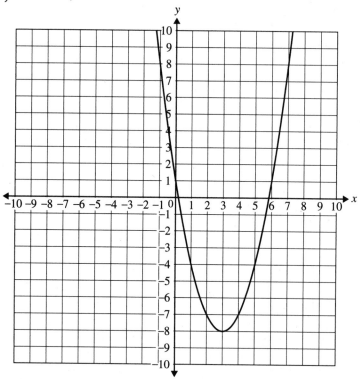

15. $y = x^2 - 4x + 3$

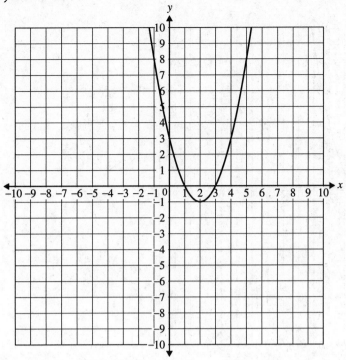

16. $y = -x^2 + 2x + 5$

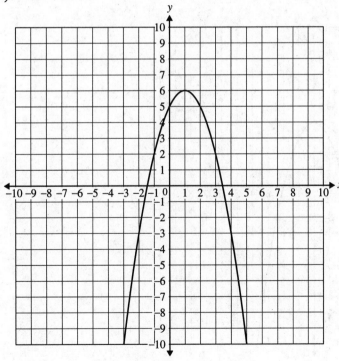

17. $y = x^2 + 6x + 9$

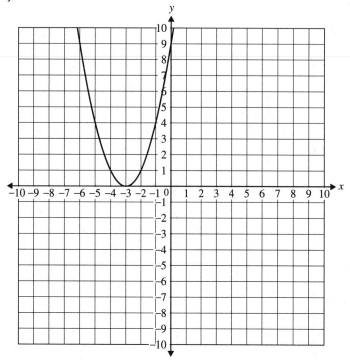

18. $y = 4 - x^2$

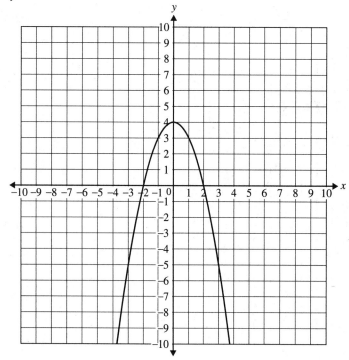

19. $y = 2x^2 + 4x - 1$

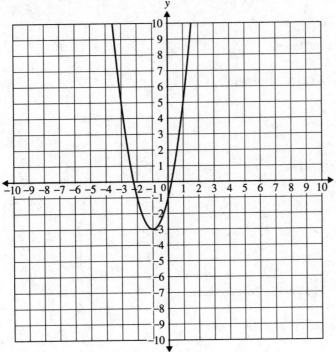

20. $y = x^2 - 9$

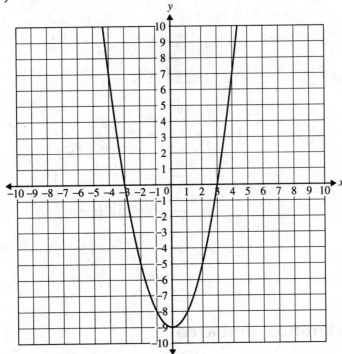

11 Polygons

11·1

1. $\triangle ABC$ is a scalene right triangle.

2. $\triangle XYZ$ is an acute isosceles triangle.

3. $\triangle RST$ is a acute scalene triangle.

4. $\triangle DOG$ is an equilateral, equiangular, acute triangle.

5. $\triangle CAT$ is an obtuse scalene triangle.

6. The measure of $\angle E$ is $180° - (35° + 35°) = 110°$.

7. The measure of $\angle B$ is $\dfrac{180° - 100°}{2} = 40°$.

8. $AC = 13$ inches

9. $ST = 7$ centimeters

10. $YZ = 6\sqrt{3} \approx 10.39$ meters

11·2

1. $\angle D = 96°$
2. $\angle C = 73°$
3. $\angle A = 115°$
4. $PQ = \dfrac{12+28}{2} = \dfrac{40}{2} = 20$ centimeters
5. $15 = \dfrac{8+CD}{2} \Rightarrow 8+CD = 30 \Rightarrow CD = 22$ inches
6. Parallelogram
7. Not a parallelogram
8. Parallelogram
9. Cannot be determined
10. $\angle B = 128°$, $\angle C = 52°$, and $\angle D = 128°$
11. $\angle B = 87°$, $\angle C = 93°$, and $\angle D = 87°$
12. $CD = 12$ centimeters
13. $AD = 9$ inches
14. Square
15. Parallelogram
16. Rectangle
17. Rhombus
18. The length of diagonal $AC = 12\sqrt{2} \approx 16.97$ centimeters.
19. $RT = 12$ inches
20. $XZ = 5$ meters
21. $MK = 24$ centimeters
22. $\angle SPR = 62°$
23. $\angle AED = 90°$
24. $\angle NJK = 45°$
25. $\angle ZWX = 74°$

11·3

1. Pentagon: $180(5-2) = 540°$
2. Octagon: $180(8-2) = 1{,}080°$
3. Hexagon: $180(6-2) = 720°$
4. Decagon: $180(10-2) = 1{,}440°$
5. 18-gon: $180(18-2) = 2{,}880°$
6. $\dfrac{180(8-2)}{8} = 135°$
7. $\dfrac{180(12-2)}{12} = 150°$
8. $\dfrac{180(6-2)}{6} = 120°$
9. $\dfrac{180(20-2)}{20} = 162°$
10. $\dfrac{180(5-2)}{5} = 108°$
11. $\dfrac{360°}{6} = 60°$
12. $360 - (70+74+82+61) = 73°$
13. $\dfrac{360°}{15} = 24°$
14. $\dfrac{8 \times (8-3)}{2} = 20$ diagonals
15. $\dfrac{9 \times (9-3)}{2} = 27$ diagonals

12 Perimeter, circumference, and area

12·1

1. 112 square centimeters
2. 81 square meters
3. 11 centimeters
4. 16 centimeters
5. $2w + 2(w+7) = 38 \Rightarrow w = 6 \Rightarrow l = 13 \Rightarrow A = 78$ square centimeters
6. $A = 216$ square centimeters
7. $A = \dfrac{1}{2} \cdot 5(3+7) = 25$ square meters
8. $A = 112$ square centimeters

9. $A = \dfrac{1}{2} \cdot 9(9 + 35) = 198$ square centimeters

10. $740 = \dfrac{1}{2} h(31 + 43) \Rightarrow 740 = 37h \Rightarrow h = 20$ centimeters

11. $A = 7.5$ square centimeters

12. $1{,}743$ square centimeters

13. $h = \sqrt{35^2 - 21^2} = 28 \Rightarrow A = \dfrac{1}{2} \cdot 42 \cdot 28 = 588$ square centimeters

14. $h = \sqrt{16^2 - 8^2} = 8\sqrt{3} \approx 13.86 \Rightarrow A = \dfrac{1}{2} \cdot 16 \cdot 8\sqrt{3} = 64\sqrt{3} \approx 110.85$ square centimeters

15. $149.5 = \dfrac{1}{2} \cdot 23 \cdot h \Rightarrow h = \dfrac{149.5}{11.5} = 13$ centimeters

12·2

1. $A = \dfrac{1}{2} \cdot 7\sqrt{3} \cdot 84 = 294\sqrt{3} \approx 509.22$ square centimeters

2. $A = \dfrac{1}{2} \cdot 12 \cdot (8 \cdot 10) = 480$ square centimeters

3. $a = \sqrt{13^2 - 5^2} = 12 \Rightarrow A = \dfrac{1}{2} \cdot 12 \cdot (5 \cdot 10) = 300$ square centimeters

4. $a = \sqrt{20^2 - 19^2} = \sqrt{39} \Rightarrow A = \dfrac{1}{2} \cdot 19 \cdot (20\sqrt{39}) = 190\sqrt{39} \approx 1{,}186.55$ square centimeters

5. $a = \sqrt{4^2 - (2\sqrt{3})^2} = 2 \Rightarrow A = \dfrac{1}{2} \cdot 2 \cdot (3 \cdot 4\sqrt{3}) = 12\sqrt{3} \approx 20.78$ square centimeters

6. $1{,}458\sqrt{3} = \dfrac{1}{2} \cdot a \cdot 108\sqrt{3} \Rightarrow a = \dfrac{1{,}458\sqrt{3}}{54\sqrt{3}} = 27$ centimeters

7. $2{,}028 = \dfrac{1}{2} \cdot a \cdot (8 \cdot 21) \Rightarrow a = \dfrac{2{,}028}{84} \approx 24$ centimeters

8. $292.5 = \dfrac{1}{2} \cdot 9 \cdot P \Rightarrow P = \dfrac{292.5}{4.5} = 65$ centimeters

9. $2{,}520 = \dfrac{1}{2} \cdot 28 \cdot P \Rightarrow P = \dfrac{2{,}520}{14} = 180 \Rightarrow s = \dfrac{180}{10} = 18$ centimeters

10. $36\sqrt{3} = \dfrac{1}{2} \cdot 2\sqrt{3} \cdot P \Rightarrow P = \dfrac{36\sqrt{3}}{\sqrt{3}} = 36 \Rightarrow s = 12$ centimeters

12·3

1. $A = 25\pi \approx 78.5$ square centimeters
2. $A = 121\pi \approx 380$ square meters
3. $A = 81\pi \approx 254$ square centimeters
4. $r = 7$ meters
5. $d = 22$ centimeters
6. $r = \sqrt{16} = 4 \Rightarrow d = 8 \Rightarrow C = 8\pi \approx 25.1$ inches
7. $A = \dfrac{40°}{360°} \cdot \pi \cdot 18^2 = 36\pi \approx 113$ square centimeters

8. $27\pi = \dfrac{120°}{360°} \cdot \pi \cdot r^2 \Rightarrow r^2 = \dfrac{27\pi}{\pi/3} = 81 \Rightarrow r = 9$ centimeters

9. $S = \dfrac{18°}{360°} \cdot \pi \cdot 28 = 1.4\pi \approx 4.4$ centimeters

10. $20\pi = \dfrac{150°}{360°} \cdot \pi \cdot d \Rightarrow d = \dfrac{20\pi}{5\pi/12} = 48 \Rightarrow r = 24$ inches

13 Transformations

1.

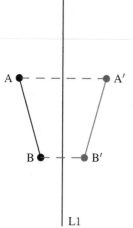

2.

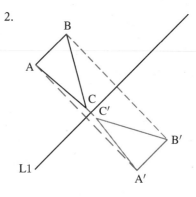

3.

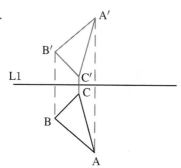

4.

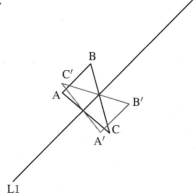

5.

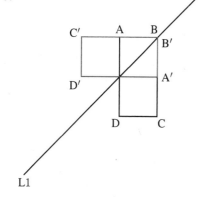

6. A′(1, 2) and B′(3, −2)

7. R′(2, −2), S′(6, 1), and T′(1, 3)

8. A′(4, 3), B′(4, 5), C′(−1, 5), and D′(−1, 3)

9. X′(−7, −2), Y′(1, −6), and Z′(−1, 1)

10. P′(10, 5), Q′(7, 5), R′(7, 2), and S′(10, 2)

1.

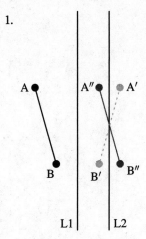

2.

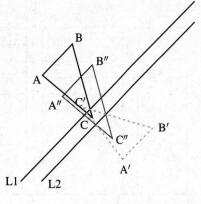

3.

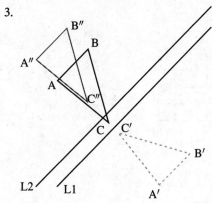

4.

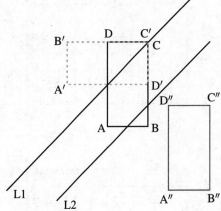

5.

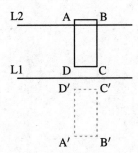

6.

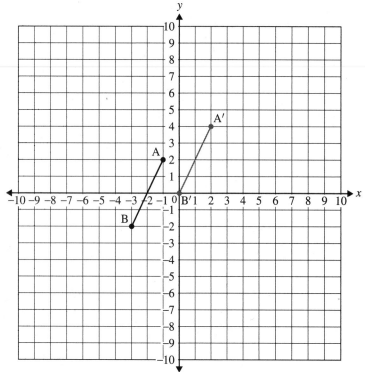

7.

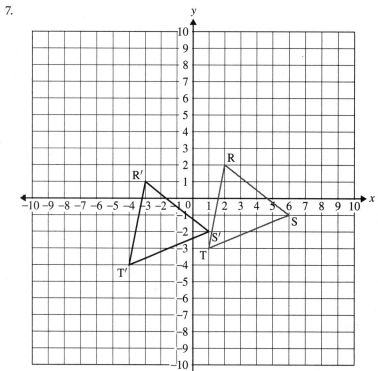

8.

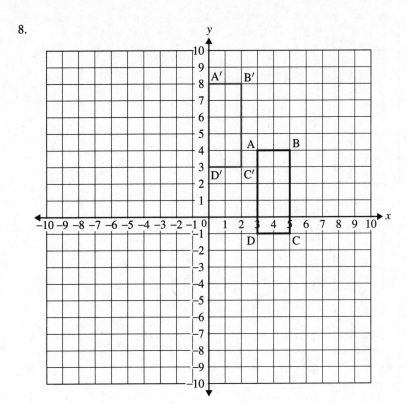

9.

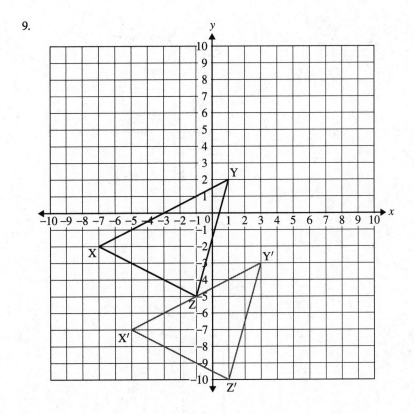

10.

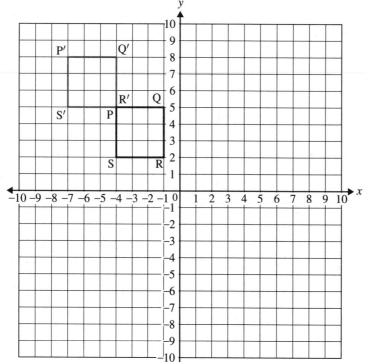

13·3

1.

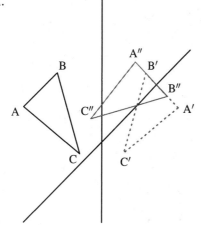

2.

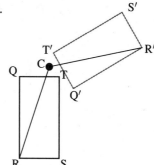

3.

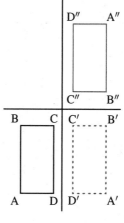

4.

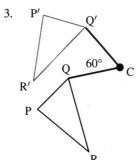

5.

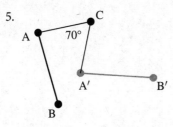

6. A′(−2, −1) and B′(2, −3)

7. R′(−2, −2), S′(−6, 1), and T′(−1, 3)

8. A′(−4, 3), B′(−4, 5), C′(1, 5), and D′(1, 3)

9. X′(2, −7), Y′(−2, 1), and Z′(5, −1)

10. P′(5, 4), Q′(5, 1), R′(2, 1), and S′(2, 4)

13·4   1.

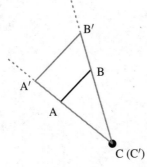

2.

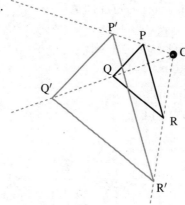

3.

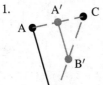

4.

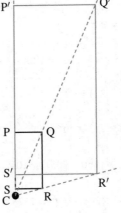

5.

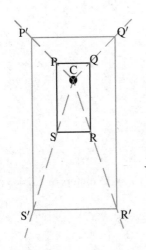

6. A'(−3, 6) and B'(−9, −6).

7. P'(−3, 9), Q'(3, 9), R'(3, 3), and S'(−3, 3)

8. A'(1.5, 2), B'(2.5, 2), C'(2.5, −0.5), and D'(1.5, −0.5)

9. A'(−2, 3) and B'(−10, −5).

10. R'(1, 3), S'(3, 1.5), and T'(0.5, 0.5)

14   Circles

14·1
1. $\overleftrightarrow{AB}$ is a secant of circle C.

2. $\overline{JK}$ is a diameter of circle O.

3. $\overline{FG}$ is a chord of circle O.

4. $\overline{CD}$ is a radius of circle C.

5. $\overrightarrow{LB}$ is a tangent of circle O.

6. Diameter: $\overline{JK}$

7. Tangent: $\overrightarrow{LB}$ or $\overleftrightarrow{EH}$

8. Radius: $\overline{CD}$ or $\overline{OJ}$ or $\overline{OK}$

9. Chord: $\overline{FG}$ or $\overline{JK}$ or $\overline{AB}$

10. Secant: $\overleftrightarrow{AB}$

14·2
1. $\angle O = 48°$

2. $\angle NPQ = 43°$

3. $\overparen{NQ} = 45°$

4. $\overparen{NQ} = 106°$

5. $\angle NPQ = 14°$

6. $\angle 1 = \dfrac{96 + 62}{2} = 79°$

7. $\angle 4 = \dfrac{101 + 93}{2} = 97°$

8. $\angle 3 = \dfrac{118 + 84}{2} = 101°$

9. $\angle 2 = \dfrac{125 + 63}{2} = 94°$

10. $\angle 4 = \dfrac{87 + 101}{2} = 94° \Rightarrow \angle 1 = 180° - 94° = 86°$

11. $\angle WZX = \dfrac{1}{2}(85 - 23) = 31°$

12. $\angle VZW = \dfrac{1}{2}(93 - 21) = 36°$

13. $\angle VZY = \dfrac{1}{2}(285 - (360 - 285)) = 105°$

14. $57° = \dfrac{1}{2}(157° - \overparen{VRT}) \Rightarrow 114° = 157° - \overparen{VRT} \Rightarrow \overparen{VRT} = 43°$

15. $33° = \dfrac{1}{2}(79° - \overparen{RT}) \Rightarrow 66° = 79° - \overparen{RT} \Rightarrow \overparen{RT} = 13°$

14·3
1. $PR = 12$ centimeters
2. $PT = 28$ inches
3. $PO = \sqrt{12^2 + 5^2} = 13$ meters
4. $OR = \sqrt{35^2 - 28^2} = 21$ feet
5. $ED = 4$ centimeters
6. $EB = 8$ inches
7. $13 \cdot 3 = x \cdot (16 - x) \Rightarrow x^2 - 16x + 39 = 0 \Rightarrow (x - 13)(x - 3) = 0 \Rightarrow CE = 13$ meters or $CE = 3$ meters
8. $10 \cdot 10 = 25 \cdot VW \Rightarrow VW = 4$ feet
9. $OV = OX = 5 \Rightarrow OZ = \sqrt{5^2 - 3^2} = 4 \Rightarrow ZW = 5 - 4 = 1$ centimeter
10. $XZ \cdot ZY = VZ \cdot ZW \Rightarrow x^2 = 8 \cdot 18 \Rightarrow x = 12 \Rightarrow XY = 24$ inches

15 Volume and surface area

15·1

1. $V = 30 \times 14 \times 8 = 3{,}360$ cubic centimeters

2. $V = \dfrac{1}{2} \times 3 \times 4 \times 4 = 24$ cubic inches

3. $V = \dfrac{1}{2} \times 8 \times 12 \times 10 = 480$ cubic inches

4. $V = 8 \times 5 \times 9 = 360$ cubic inches

5. $V = 32 \times 15 = 480$ cubic centimeters

6. $V = 85 \times 12 = 1{,}020$ cubic inches

7. $V = 1{,}216\pi \approx 3{,}818.24$ cubic centimeters

8. $2{,}025\pi \approx 6{,}358.5$ cubic inches

9. $d = 2r \Rightarrow r = \dfrac{12}{2} = 6$ then $V = 324\pi \approx 1{,}017.36$ cubic centimeters

10. $C = 2\pi r \Rightarrow r = \dfrac{119.32}{2 \times 3.14} = 19$ then $V = 9{,}025\pi \approx 28{,}338.5$ cubic feet

15·2

1. $V = \dfrac{1}{3} \times 15^2 \times 18 = 1{,}350$ cubic inches

2. $V = \dfrac{1}{3} \times \dfrac{1}{2} \times 8 \times 4\sqrt{3} \times 9 \approx 83.14$ cubic centimeters

3. $V = \dfrac{1}{3} \times 7 \times 9 \times 5 = 105$ cubic feet

4. $V = \dfrac{1}{3} \times 150 \times 12 = 600$ cubic centimeters

5. $V = \dfrac{1}{3} \times 24^2 \times 24 = 4{,}608$ cubic feet

6. $V = \dfrac{1}{3} \times \pi \times 12^2 \times 8 = 384\pi \approx 1{,}205.76$ cubic inches

7. $V = \dfrac{1}{3} \times \pi \times 8^2 \times 12 = 256\pi \approx 803.84$ cubic inches

8. $d = 2r \Rightarrow r = \dfrac{16}{2} = 8$ then $V = \dfrac{1}{3} \times \pi \times 8^2 \times 9 = 192\pi \approx 602.88$ cubic centimeters

9. $C = 2\pi r \Rightarrow r = \dfrac{56.52}{2 \times 3.14} = 9$ then $V = \dfrac{1}{3} \times \pi \times 9^2 \times 15 = 405\pi \approx 1{,}271.7$ cubic meters

10. Pyramid: $V = \dfrac{1}{3} \times 6^2 \times 8 = 96$ cubic feet. Cone: $V = \dfrac{1}{3} \times \pi \times (3.82)^2 \times 8 \approx 122.19$ cubic feet.

The cone has the larger volume.

15·3

1. $V = \dfrac{4}{3}\pi \cdot 3^3 = 36\pi \approx 113.04$ cubic feet

2. $V = \dfrac{4}{3}\pi \cdot 12^3 = 2{,}304\pi \approx 7{,}234.56$ cubic centimeters

3. $V = \dfrac{4}{3}\pi \cdot 27^3 = 26{,}244\pi \approx 82{,}406.16$ cubic meters

4. $V = \dfrac{4}{3}\pi \cdot 6^3 = 288\pi \approx 904.32$ cubic inches

5. $V = \dfrac{4}{3}\pi \cdot 30^3 = 36{,}000\pi \approx 113{,}040$ cubic centimeters

6. $r = 15$ feet, $C = 30\pi \approx 94.2$ feet, $A = 225\pi \approx 706.5$ square feet, $V = 4{,}500\pi \approx 14{,}580$ cubic feet

7. $r = 48$ centimeters, $C = 96\pi \approx 301.44$ centimeters, $A = 2304\pi \approx 7{,}234.56$ square centimeters, $V = 147{,}456\pi \approx 477{,}757.4$ cubic centimeters

8. $r = 2$ meters, $C = 4\pi \approx 12.56$ meters, $A = 4\pi \approx 12.56$ square meters, $V = 10\dfrac{2}{3}\pi \approx 34.56$ cubic meters

9. Cone: $V = \frac{1}{3} \pi \cdot 1^2 \cdot 6 = 2\pi \approx 6.28$ cubic inches. Sphere: $V = \frac{4}{3} \pi r^3 \Rightarrow 2\pi = \frac{4}{3} \pi r^3 \Rightarrow 2 = \frac{4}{3} r^3 \Rightarrow \frac{6}{4} = \frac{3}{2} = r^3$.

   The largest spherical scoop of ice cream is $\sqrt[3]{1.5} \approx 1.14$ inches.

10. Unisphere: $V = \frac{4}{3} \pi \cdot 60^3 = 288{,}000\pi \approx 904{,}320$ cubic feet. Model: $V = \frac{4}{3} \pi \cdot 20^3 = 10{,}666\frac{2}{3}\pi \approx 33{,}493\frac{1}{3}$

    cubic inches. 1 cubic foot $= 12^3$ cubic inches, so the model is $33{,}493\frac{1}{3} \div 12^3 \approx 19.38$ cubic feet.

    $\dfrac{19.38}{904{,}320} = \dfrac{1{,}938}{90{,}432{,}000} \approx \dfrac{1}{46{,}663}$

15·4

1. $S = 2 \cdot 18 \cdot 10 + 2 \cdot 18 \cdot 9 + 2 \cdot 10 \cdot 9 = 360 + 324 + 180 = 864$ square inches

2. $S = 2 \cdot \frac{1}{2} \cdot 3 \cdot 4 + (3 + 4 + 5) \cdot 2 = 12 + 24 = 36$ square inches

3. $S = 2(84.3) + 35 \cdot 7 = 168.6 + 245 = 413.6$ square centimeters

4. $S = 2 \cdot \left(\frac{1}{2} \cdot 4 \cdot 2\sqrt{3} \right) + (3 \cdot 4) \cdot 8 = 8\sqrt{3} + 96 \approx 109.86$ square inches

5. $S = 6 \cdot 11^2 = 6 \cdot 121 = 726$ square inches

6. $S = 2\pi \cdot 2^2 + 2\pi \cdot 2 \cdot 9 = 8\pi + 36\pi = 44\pi \approx 138.16$ square inches

7. $d = 12 \Rightarrow r = 6$ then $S = 2\pi \cdot 6^2 + 2\pi \cdot 6 \cdot 15 = 72\pi + 180\pi = 252\pi \approx 791.28$ square centimeters

8. $C = 314 \Rightarrow d = 100 \Rightarrow r = 50$ then $S = 2\pi \cdot 50^2 + 2\pi \cdot 50 \cdot 14 = 5{,}000\pi + 1{,}400\pi = 6{,}400\pi \approx$ 20,096 square inches

9. $S = 2lw + 2lh + 2wh$

   $2{,}512 = 2 \cdot 40 \cdot 16 + 2 \cdot 40 \cdot h + 2 \cdot 16 \cdot h$

   $2{,}512 = 1{,}280 + 80h + 32h$

   $2{,}512 = 1{,}280 + 112h$

   $1{,}232 = 112h$

   $h = 11$

10. $S = 2\pi r^2 + 2\pi rh$

    $357.96 = 2 \cdot \pi \cdot 6^2 + 2 \cdot \pi \cdot 6 \cdot h$

    $357.96 = 72\pi + 12\pi \cdot h$

    $357.96 = 72(3.14) + 12(3.14)h$

    $357.96 = 226.08 + 37.68h$

    $131.88 = 37.68h$

    $h = 3.5$

15·5

1. $S = 24^2 + \frac{1}{2}(4 \cdot 24) \cdot 13 = 576 + 624 = 1{,}200$ square feet

2. $S = \left(\frac{1}{2} \cdot 12 \cdot 6\sqrt{3} \right) + \frac{1}{2}(3 \cdot 12) \cdot 10 \approx 62.35 + 180 \approx 242.35$ square centimeters

3. $S = 509.22 + \frac{1}{2} \cdot 84 \cdot 16 = 509.22 + 672 = 1{,}181.22$ square inches

4. $S = 480 + \frac{1}{2} \cdot 192 \cdot 30 = 480 + 2{,}880 = 3{,}360$ square meters

5. $S = \pi r^2 + \pi rl$

   $= \pi \cdot 25^2 + \pi \cdot 25 \cdot 32$

   $= 625\pi + 800\pi = 1425\pi$

   $\approx 1{,}962.5 + 2{,}512 \approx 4{,}474.5$ square centimeters

6. $S = \pi r^2 + \pi r l$

$\quad = \pi \cdot 12^2 + \pi \cdot 12 \cdot 24$

$\quad = 144\pi + 288\pi = 432\pi$

$\quad \approx 452.16 + 904.32 \approx 1,356.48$ square inches

7. $S = \pi r^2 + \pi r l$

$\quad = \pi \cdot 11^2 + \pi \cdot 11 \cdot 19$

$\quad = 121\pi + 209\pi = 330\pi$

$\quad \approx 379.94 + 656.26 \approx 1,036.2$ square centimeters

8. $C = 201 \Rightarrow r \approx 32$ then $S = \pi r^2 + \pi r l$

$\quad = \pi \cdot 32^2 + \pi \cdot 32 \cdot 56$

$\quad = 1,024\pi + 1,792\pi = 2,816\pi$

$\quad \approx 3,215.36 + 5,626.88 \approx 8,842.24$ square inches

9. $\qquad S = 139.36 + P \cdot 11$

$\quad 1,184.36 = 139.36 + P \cdot 11$

$\qquad 1045 = 11P$

$\qquad\quad P = 95$ meters

Because the pentagon is regular, each side is $95 \div 5$, or 19 meters.

10. $\qquad S = \pi \cdot 5^2 + \pi \cdot 5 \cdot l$

$\quad 282.75 = 25\pi + 5\pi l$

$\quad 282.75 = 78.5 + 15.7l$

$\quad 204.25 = 15.7l$

$\qquad\quad l \approx 13$ inches

15·6

1. $S = 4\pi \cdot 2^2 = 16\pi \approx 50.24$ square inches
2. $S = 4\pi \cdot 5^2 = 100\pi \approx 314$ square centimeters
3. $S = 4\pi \cdot 8^2 = 256\pi \approx 803.84$ square meters
4. $S = 4\pi \cdot 12^2 = 576\pi \approx 1,808.64$ square feet
5. $S = 4\pi \cdot 21^2 = 1,764\pi \approx 5,538.96$ square yards

6. $S = 4\pi \cdot 18^2 = 1,296\pi \approx 4,069.44$ square inches
7. $S = 4\pi \cdot 25^2 = 2,500\pi \approx 7,850$ square millimeters
8. $S = 36\pi = 4\pi r^2 \Rightarrow r^2 = 9 \Rightarrow r = 3$ meters
9. $S = 64\pi = 4\pi r^2 \Rightarrow r^2 = 16 \Rightarrow r = 4$ inches
10. $S = 144\pi = 4\pi r^2 \Rightarrow r^2 = 36 \Rightarrow r = 6$ centimeters

16 Counting and probability

16·1

1. $2 \times 3 \times 4 = 24$
2. $2 \times 3 \times 5 = 30$
3. $8 \times 12 \times 4 = 384$
4. $26 \times 26 \times 10 \times 10 \times 10 \times 10 \times 10 = 67,600,000$
5. $26 \times 26 \times 10 \times 9 \times 8 \times 7 \times 6 = 20,442,240$
6. $26 \times 26 \times 26 \times 26 \times 26 \times 26 = 308,915,776$
7. $10 \times 10 \times 10 \times 10 \times 10 \times 10 = 1,000,000$
8. $36 \times 36 \times 36 \times 36 \times 36 \times 36 = 2,176,782,336$
9. a) $10 \times 10 \times 10 = 1,000$
    b) $10 \times 10 = 100$
    c) $10 \times 10 \times 10 \times 10 = 10,000$
    d) $1,000 \times 100 \times 10,000 = 1,000,000,000$
10. This restriction doesn't change the two-digit or the four-digit section, but it eliminates 102 of the three-digit possibilities, leaving $898 \times 100 \times 10,000 = 898,000,000$ possible numbers.
11. $1 \times 4 \times 8 = 32$
12. $1 \times 8 \times 7 \times 6 = 336$
13. $2 \times 4 \times 3 \times 8 = 192$
14. $2 \times 4 \times 8 \times 7 = 448$
15. $2 \times 4 \times 3 \times 2 \times 8 \times 7 \times 6 \times 5 = 80,640$

16·2

1. $\dfrac{3!}{2!} = \dfrac{3 \times 2 \times 1}{2 \times 1} = 3$

2. $\dfrac{5!}{3!} + \dfrac{6!}{4!} = \dfrac{5 \times 4 \times \cancel{3 \times 2 \times 1}}{\cancel{3 \times 2 \times 1}} + \dfrac{6 \times 5 \times \cancel{4 \times 3 \times 2 \times 1}}{\cancel{4 \times 3 \times 2 \times 1}} = 20 + 30 = 50$

3. $8! - 4! = 8 \times 7 \times 6 \times 5 \times 4 \times 3 \times 2 \times 1 - 4 \times 3 \times 2 \times 1 = 40{,}320 - 24 = 40{,}296$

4. $\dfrac{8! \times 5!}{6!} = \dfrac{8 \times 7 \times \cancel{6 \times 5 \times 4 \times 3 \times 2 \times 1} \times 5 \times 4 \times 3 \times 2 \times 1}{\cancel{6 \times 5 \times 4 \times 3 \times 2 \times 1}} = 6{,}720$

5. $\dfrac{6! \times 4!}{5! \times 7!} = \dfrac{6 \times \cancel{5 \times 4 \times 3 \times 2 \times 1} \times \cancel{4 \times 3 \times 2 \times 1}}{\cancel{5 \times 4 \times 3 \times 2 \times 1} \times 7 \times 6 \times 5 \times \cancel{4 \times 3 \times 2 \times 1}} = \dfrac{6}{7 \times 6 \times 5} = \dfrac{1}{35}$

6. $_8P_5 = \dfrac{8!}{(8-5)!} = 8 \times 7 \times 6 \times 5 \times 4 = 6{,}720$

7. $_7P_4 = \dfrac{7!}{3!} = 7 \times 6 \times 5 \times 4 = 840$

8. $_9P_8 = \dfrac{9!}{1!} = 362{,}880$

9. $_{10}P_2 = 90$

10. $_{12}P_1 = 12$

11. $_6P_4 = 360$

12. $_6P_5 = 720$

13. $_7P_3 = 210$

14. $_7P_4 = 840$

15. $_{10}P_6 = 151{,}200$

16·3

1. $_6C_3 = 20$

2. $_9C_7 = 36$

3. $_{12}C_5 = 792$

4. $_5C_5 = 1$

5. $_5C_1 = 5$

6. $_{40}C_4 = 91{,}390$

7. $_{52}C_5 = 2{,}598{,}960$

8. $_{52}C_{13} \approx 6.35 \times 10^{11}$

9. $_{56}C_5 = 3{,}819{,}816$

10. $_{56}C_6 = 32{,}468{,}436$

16·4

1. $\dfrac{_8P_8}{2!} = \dfrac{40{,}320}{2} = 20{,}160$

2. $\dfrac{_{10}P_9}{2! \times 3!} = \dfrac{3{,}628{,}800}{2 \times 6} = 302{,}400$

3. $\dfrac{_{12}P_7}{2! \times 2! \times 2!} = \dfrac{3{,}991{,}680}{8} = 498{,}960$

4. $\dfrac{_9P_5}{2!} = \dfrac{15{,}120}{2} = 7{,}560$

5. $\dfrac{_8P_6}{3! \times 2!} = \dfrac{20{,}160}{12} = 1{,}680$

6. $\dfrac{_6P_4}{2!} = \dfrac{360}{2} = 180$

7. $\dfrac{_8P_8}{2!} = \dfrac{40{,}320}{2} = 20{,}160$

8. $\dfrac{_{11}P_{10}}{4! \times 4! \times 2!} = \dfrac{39{,}916{,}800}{1152} = 34{,}650$

9. $\dfrac{_{10}P_5}{2! \times 2!} = \dfrac{30{,}240}{4} = 7{,}560$

10. $\dfrac{_{11}P_4}{2!} = \dfrac{7{,}920}{2} = 3{,}960$

16·5

1. $\dfrac{26}{52} = \dfrac{1}{2}$

2. $\dfrac{13}{52} = \dfrac{1}{4}$

3. $\dfrac{16}{52} = \dfrac{4}{13}$

4. $\dfrac{1}{6}$

5. $\dfrac{3}{6} = \dfrac{1}{2}$

6. $\dfrac{6}{6} = 1$

7. $\dfrac{0}{6} = 0$

8. $\dfrac{5}{25} = \dfrac{1}{5}$

9. $\dfrac{15}{25} = \dfrac{3}{5}$ or $1 - \dfrac{10}{15}$

10. $\dfrac{0}{25} = 0$

16·6

1. $\dfrac{26}{52} \times \dfrac{4}{52} = \dfrac{1}{2} \times \dfrac{1}{13} = \dfrac{1}{26}$

2. $\dfrac{26}{52} \times \dfrac{12}{52} = \dfrac{1}{2} \times \dfrac{3}{13} = \dfrac{3}{26}$

3. $\dfrac{13}{52} \times \dfrac{16}{52} = \dfrac{1}{4} \times \dfrac{4}{13} = \dfrac{1}{13}$

4. $\dfrac{13}{52} \times \dfrac{36}{52} = \dfrac{1}{4} \times \dfrac{9}{13} = \dfrac{9}{52}$

5. $\dfrac{1}{6} \times \dfrac{1}{6} = \dfrac{1}{36}$

6. $\dfrac{1}{6} \times \dfrac{3}{6} = \dfrac{3}{36} = \dfrac{1}{12}$

7. $\dfrac{1}{6} \times \dfrac{1}{6} = \dfrac{1}{36}$

8. $\dfrac{3}{6} \times \dfrac{3}{6} = \dfrac{1}{2} \times \dfrac{1}{2} = \dfrac{1}{4}$

9. The only way to get a total of 2 is a 1 on each die. $\dfrac{1}{6} \times \dfrac{1}{6} = \dfrac{1}{36}$

10. You can get a 7 with any number on the red die, if you get the right number on the white die. $\dfrac{6}{6} \times \dfrac{1}{6} = \dfrac{1}{6}$

16·7

1. Mutually exclusive $\dfrac{4}{52} + \dfrac{4}{52} = \dfrac{1}{13} + \dfrac{1}{13} = \dfrac{2}{13}$

2. Mutually exclusive $\dfrac{16}{52} + \dfrac{4}{52} = \dfrac{4}{13} + \dfrac{1}{13} = \dfrac{5}{13}$

3. Not mutually exclusive $\dfrac{13}{52} + \dfrac{4}{52} - \dfrac{1}{52} = \dfrac{16}{52} = \dfrac{4}{13}$

4. Not mutually exclusive $\dfrac{26}{52} + \dfrac{12}{52} - \dfrac{6}{52} = \dfrac{32}{52} = \dfrac{8}{13}$

5. $\dfrac{6}{24} + \dfrac{4}{24} = \dfrac{10}{24} = \dfrac{5}{12}$

6. $\dfrac{2}{24} + \dfrac{4}{24} = \dfrac{6}{24} = \dfrac{1}{4}$

7. $\dfrac{9}{24} + \dfrac{3}{24} = \dfrac{12}{24} = \dfrac{1}{2}$

8. $\dfrac{2}{24} + \dfrac{6}{24} = \dfrac{8}{24} = \dfrac{1}{3}$

9. $\dfrac{6}{24} + \dfrac{0}{24} = \dfrac{6}{24} = \dfrac{1}{4}$

10. $\left(\dfrac{4}{52} \times \dfrac{13}{52}\right) + \left(\dfrac{4}{52} \times \dfrac{13}{52}\right) = \dfrac{1}{52} + \dfrac{1}{52}$

11. $\left(\dfrac{4}{52} + \dfrac{4}{52}\right) \times \left(1 - \dfrac{13}{52}\right) = \dfrac{8}{52} \times \dfrac{39}{52} = \dfrac{2}{13} \times \dfrac{3}{4} = \dfrac{3}{26}$

12. $1 - \left(\dfrac{26}{52} \times \dfrac{4}{52}\right) = 1 - \left(\dfrac{1}{2} \times \dfrac{1}{13}\right) = 1 - \dfrac{1}{26} = \dfrac{25}{26}$

13. $1 - \left(\dfrac{4}{52} + \dfrac{4}{52}\right) = 1 - \dfrac{8}{52} = \dfrac{44}{52} = \dfrac{11}{13}$

14. $\dfrac{26}{52} \times \left(1 - \dfrac{12}{52}\right) = \dfrac{1}{2} \times \dfrac{40}{52} = \dfrac{1}{2} \times \dfrac{10}{13} = \dfrac{5}{13}$

15. $\left(1 - \left(\dfrac{4}{52} + \dfrac{4}{52}\right)\right) \times \dfrac{26}{52} = \left(1 - \dfrac{8}{52}\right) \times \dfrac{1}{2} = \left(1 - \dfrac{2}{13}\right) \times \dfrac{1}{2} = \dfrac{11}{13} \times \dfrac{1}{2} = \dfrac{11}{26}$

1. $\dfrac{750}{1335} = \dfrac{50}{89}$

2. $\dfrac{625}{1375} = \dfrac{5}{11}$

3. $\dfrac{115}{1590} = \dfrac{23}{318}$

4. $\dfrac{750}{1375} = \dfrac{6}{11}$

5. $\dfrac{115}{200} = \dfrac{23}{40}$

6. $\dfrac{3}{51} = \dfrac{1}{17}$

7. $\dfrac{25}{51}$

8. $\dfrac{11}{51}$

9. $\dfrac{12}{51} = \dfrac{4}{17}$

10. $\dfrac{25}{51}$

17 Statistics

1. Mode = 28
2. Median = 74
3. Mean = 135
4. The most frequent response was Chocolate Fudge (with a frequency of 938,764,402).
5. Mode = 27
6. Median = 27
7. Mean ≈ 27.5
8. The most common interval was greater than 10 but less than or equal to 20 innings pitched.
9. The median number of innings pitched is approximately 20 innings. Nine pitchers pitched less than or equal to 20 innings, and 9 pitched more than 20 innings.
10.

Interval	Class mark	# of pitchers	product
0–0	5	4	20
10–20	15	5	75
20–30	25	4	100
30–40	35	0	0
40–50	45	0	0
50–60	55	0	0
60–70	65	2	130
70–80	75	1	75
80–90	85	1	85
90–100	95	0	0
100–110	105	1	105
Totals:		18	590
Mean = 590 ÷ 18 ≈ 32.8			

11. Median = $\dfrac{\$25{,}000 + \$30{,}000}{2} = \$27{,}500$

12. Mean = \$38,250
13. Median = \$27,500
14. Mean = \$48,250
15. The choice of the median is based on its position in the ordered data. Raising the top salary doesn't change the position of the middle values. The mean adds values and divides by the number of values. The increase in the top salary makes the total larger, but that total is divided by the same number. As a result the mean is higher.
16. Median = \$37,500
17. Mean = \$48,250
18. Both the median and the mean would increase by \$5,000, so median = \$42,500 and mean = \$53,250.
19. If the average for 3 rounds was 90, the total, before dividing by 3, must have been 3 × 90 = 270. The known scores of 90 and 92 account for 182 of the 270, leaving 270 − 182 = 88. His score for the third round was 88.
20. If the mean score for 4 tests is 88, the total of the 4 tests is 4 × 88 = 352. To earn an average of 90 for 5 tests, she needs a total of 5 × 90 = 450, or a score of 450 − 352 = 98 on the fifth test.

17·2
1. Range $= 99 - 42 = 57$
2. Q1 $= 67$ and Q3 $= 87$
3. Interquartile range $= 87 - 67 = 20$
4. Q1 $= 23$ and Q3 $= 31$
5. IQR $= 31 - 23 = 8$
6. Range $= 42 - 17 = 25$
7. Range $= 52 - 19 = 33$
8. Median $= 31$
9. Q1 $= 26$ and Q3 $= 36$
10. IQR $= 36 - 26 = 10$
11. The marriage ages for men have a larger range and a larger interquartile range.
12. For women, mean plus one standard deviation $= 27.5 + 5.8 = 33.3$, mean plus two standard deviations $= 27.5 + 5.8 + 5.8 = 39.1$, and mean plus three standard deviations $= 27.5 + 5.8 + 5.8 + 5.8 = 44.9$. A woman reporting a marriage age of 45 would fall just above three standard deviations above the mean. For men, mean plus one standard deviation $= 31.9 + 7.9 = 39.8$, mean plus two standard deviations $= 31.9 + 7.9 + 7.9 = 47.7$, and mean plus three standard deviations $= 31.9 + 7.9 + 7.9 + 7.9 = 55.6$. A man reporting a marriage age of 45 would fall between one and two standard deviations above the mean. A woman who reported her age at time of marriage as 45 would be significantly above average.
13. The mean salary at ZYX Widgets is $54,000.

Salaries	Minus the mean	Squared
$20,000	−34,000	1,156,000,000
$22,000	−32,000	1,024,000,000
$30,000	−24,000	576,000,000
$35,000	−19,000	361,000,000
$50,000	−4,000	16,000,000
$55,000	1,000	1,000,000
$90,000	36,000	1,296,000,000
$130,000	76,000	5,776,000,000
Total:		10,206,000,000
$10,206,000,000 \div 7 = 1,458,000,000$		
Standard deviation $= \sqrt{1,458,000,000} \approx \$38,183.77$		

14. Q1 $= \$26,000$ and Q3 $= \$72,500$, so IQR $= \$72,500 - \$26,000 = \$46,500$
15. Range $= \$130,000 - \$20,000 = \$110,000$
16. Range $= 932 - 16 = 916$
17. Q1 $= 45$ and Q3 $= 153$
18. IQR $= 153 - 45 = 108$
19. The outlier value of 932, so much higher than the other values, pulls the mean up.
20. On a test with a mean of 75 and a standard deviation of 10, a score of 85 is one standard deviation above the mean. On a test with a mean of 78 and a standard deviation of 3, a score of 85 is more than two standard deviations above the mean, and so a better performance.

17·3
1. D. As the first variable increases, the second first decreases then increases.
2. B. As the first variable increases, the second decreases.
3. A. As the first variable increases, the second increases.
4. A. As the first variable increases, the second increases.
5. C. As the first variable increases, the second first increases then decreases.
6. Linear
7. Quadratic
8. Exponential
9. Linear
10. Quadratic
11. $y \approx 7.86x + 6.14$
12. $y = 7.2x - 0.6$

13. $y = 11x - 17.2$
14. $y = -7.6x + 78.2$
15. $y = 2.5x + 39.5$

For questions 16 through 20, answers will vary. Samples are given.

16. $y = 8x + 8$

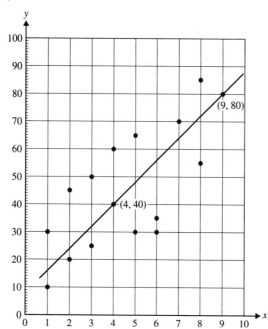

17. $y = -8\frac{1}{3}x + 98\frac{1}{3}$

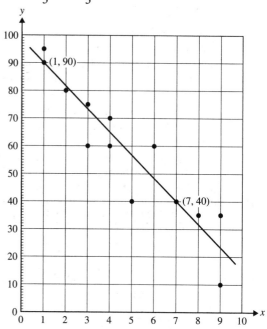

18. $y = 5x + 15$

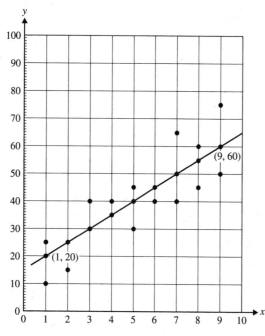

19. $y \approx -7.86x + 95.71$

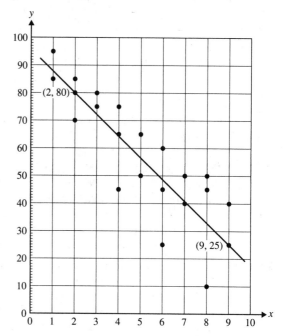

20. $y = 8.75x - 3.75$

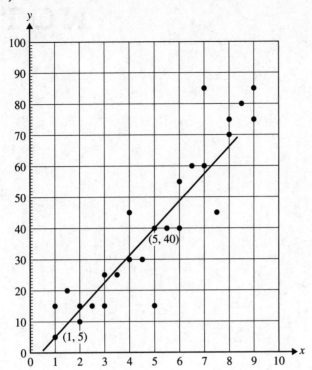

NOTES

NOTES